Getting Smar[t]
Not What Yo[u...]

Five Essentials to reach your full potential in any endeavor

By

J. B. Mason

Copyrighted Material

Getting Smarter - It's Not What You Think

Copyright © 2017 by J. B. Mason

ALL RIGHTS RESERVED.

No part of this publication may be reproduced, stored in a retrieval system, transmitted or used in any manner whatsoever without the express written permission of the publisher except for the use of brief quotations in a book review or scholarly journal.

For information about this title or to order other books and/or electronic media, contact the publisher:

Move To Excel

movetoexcel.com

ISBN-13: 978-1546561927

ISBN-10: 1546561927

Printed in the United States of America

Cover by: J. B. Mason

Human figure drawing on cover by: Andrew Cerrona

Disclaimer:

All forms of movement carry some risk. Readers are advised to take full responsibility for their safety and limits. The practices in this book is in no way intended to be a substitute for advice provided by your doctor or another medical professional. The publisher and author disclaim all liability for damage resulting from doing the practices in this book.

ACKNOWLEDGMENTS

Special thanks to my good friend Anne for her amazing support and help. Thanks to my mother whose steady encouragement has helped me through the times of doubt and worry. My brother Nick has been a stalwart believer in the approach and a grateful recipient.

Thanks to Zoe Harrington for her fine artwork on the box body and the cylinder body.

*Baby photo of Figure 20 courtesy of Irene Lyons who made the 'Baby Liv' video part of the commons.

And thanks to all former and current clients who are the inspiration I need to improve and refine the approach to be more effective every day.

Table of Contents

Introduction	**5**
1 - Why How You Move Matters	**8**
2 - Feel Your Movements	**15**
Movement Awareness Guidelines	*24*
How to Feel Practice	*25*
3 - Responding to what you feel	**28**
4 - Element 1: Breathing	**40**
5 - Element 2: Yield	**49**
External Yield	*50*
Yield from Within	*53*
Structural Connection	*56*
6 - Element 3: Differentiation	**64**
7 - Element 4: Rotation	**91**
8 - Element 5: Propulsion	**110**
9 - Putting it all Together	**117**
10 - Eliminate Pain	**126**
11 - Brain Injury & Stroke Recovery	**131**
12 - Optimize Childhood Learning	**136**
Children with special needs	*137*
13 - Embodied Spirituality	**141**
End Notes and Links	**146**

Introduction

I used to treat my body like a car: give it good quality fuel, use it regularly in the form of exercise, and fix it when parts break or wear out. I expected the body to perform as I consciously willed it, except that I had a certain level of frustration when it disappointed my expectations. The main job of the body, I thought, was to carry my brain from point A to point B so it can do the important work of thinking. I was resigned to the idea that my body wears out with age, becoming ever more feeble and fragile. My body mostly cooperated with how I thought it should function until one day it did not.

That day I woke up in pain in my right shoulder and my life was turned upside down. The doctor said I had not broken or torn anything so there was no surgical fix. He labelled the problem a neuromuscular strain and recommended drugs for the pain and physical therapy to strengthen the shoulder muscles. When neither drugs or physical therapy relieved the pain the doctor informed me I wasn't improving because the pain was "all in my head" and I needed psychological help. After that I tried alternative modalities including massage, chiropractic, and acupuncture to relieve the pain. When alternatives didn't help, I became increasingly desperate as the pain was so pervasive I couldn't sleep, concentrate at work, or find any way to relieve it. After six weeks of misery, by chance I met a guy who was attending a training in the method originated by physicist Moshe Feldenkrais[1]. After listening to my story, he said he thought the trainer might be able to help. In one session with the trainer I felt great relief from the pain and knew I was on the way to resolving it. Determined to never feel so helpless about my body again, I immediately signed up to attend a training in the method.

The training upended my notion of the body as separate from and subordinate to the mind. It taught that all information entering the brain is through our movements, and that the quality of *how we move is the source of our intelligence.* After graduating from the training in 1994, I started a part-time practice in addition to a full-time career as a systems architect in the IT industry.

The part-time practice, as taught by the training, was floundering as clients and students alike struggled to understand the importance of movement quality to their lives. I could not compete with the choice of doing a fitness class versus a movement class that felt good but wasn't considered exercise. I only saw individuals that wanted to get out of pain or that wanted to improve their technique in a physical skill. I needed to revise the approach to be more practical and easy to integrate into all functionality so that anyone would be inspired to try it.

The classes became more interactive as we focused on activities that most people struggle with such as walking up stairs, running, walking or running up or downhill, sitting, and coming from sitting to standing, and so on. In personal sessions clients are asked to move in their habitual way first and then I show them how moving differently improves function and relieves pain or discomfort. Visualization and physics, gravity, and Newton's third law of motion factor in to help people better understand how we evolved to move. For instance, visualizing a linked chain down the center of your torso is useful in helping to initiate a turn to look around yourself. As I continue to refine the approach clients are getting out of pain much quicker and improving overall functionality. Clients that had given up activities such as running, biking, hiking, skiing, tennis, golf and more find they are once again able to do those activities, and they are performing them better than ever. They are often surprised how other aspects of their lives improve, such as thinking more clearly, being less reactionary and emotional, having greater awareness of the world and their impact on it, feeling more compassion and empathy for others, and being more present and aware.

I stopped consulting in 2002 to focus full-time helping clients transform to be more functional in every activity. What every adult has in common is that certain foundational functionality, in varying degrees, has been lost. The lost functionality is essential to optimal function. Once gone few of us have any idea how to reconnect and integrate it back into how we learn or refine new skills.

I have spent the last fifteen years identifying five essential elements that are the foundation underlying all function. This book is a practical guide that shows you how to re-integrate the five foundational elements into all functionality.

You can address just your physical movements to enhance skills, get out of pain, or recover from injury. You can also choose to go far beyond the physical into a much deeper exploration of the mind body connection. Refinement of the quality of how you function leads to your body and mind converging to operate in seamless synchronicity. Through the quality of how you function you are poised to enter that special state of being where you are in timeless, effortless effort. In that state is when you can reach your greatest potential.

1 - Why How You Move Matters

"Consciousness is only possible through change; change is only possible through movement." ~ Aldous Huxley, The Art of Seeing

How often are you in a state of effortless effort as you go about your day? People call this state the zone, and to get in it they often say, "I got out of my way, I stopped thinking." The zone is defined as the state of being where you function optimally, your body and mind are inseparable, any activity you do is effortless and timeless, and you move without conscious oversight. In the zone you function non-consciously, and the mind body connection is a seamless whole working in unity and synchronicity.

Free solo rock climber Alex Honnold makes scaling impossible rock faces without ropes or safety gear look effortless.[2] He says climbing is when he is most at ease, because for him it is a form of meditation. People who meditate are trying to achieve a state of being where they are fully present and aware. *Being in a meditative state and the zone are the same thing*—all you feel, do, and think is effortless and timeless.

Babies function naturally in the zone as they learn skills such as lifting the head, rolling over, sitting up, crawling, standing, and walking. If you are fortunate enough to be around babies and toddlers, observe how they learn complex movements. You can also watch a video[3] of a baby learning to roll over. Notice how she becomes more alert, agile, and engaged as she refines the movement. Soon she is doing it effortlessly, which naturally leads to the next challenge of learning how to crawl. Children naturally feel how to refine and improve skills non-consciously, through movements. Most people that are highly gifted in a skill start learning it at a very young age. If a skill is compromised because of an injury, it can be more difficult to recapture the same level of expertise as few adults retain their natural, non-conscious learning ability. Most of us struggle to refine or learn new skills because we have stopped learning non-consciously. What happens in the transition from childhood to becoming an adult that affects how we learn?

Humans have been around for 200,000 years. For the first 99% of our existence we were hunter gatherers relying on our senses and how we moved to stay alive and procreate. Harsh climate conditions stabilized about 10,000 years ago and soon thereafter we discovered farming, allowing us to settle and form communities. We understood the mind body as a unified whole until in the seventeenth century philosopher Rene Descartes proposed that the body is essentially a machine which is controlled by the conscious mind.[4] He dismissed 200,0000 years of evolution with a single conceptual idea. His argument of the machine body being subordinate to the superior mind gained credence as society entered the industrial age. Science went one step further about seventy years ago to claim the brain functions like a computer and instructs the body what to do. The computer mind, machine body belief is now fully entrenched in society by how we educate, our approach to exercise, how we practice medicine, how we do business, how we learn new skills, and how we treat the planet. The way we function as a species is having serious global consequences that is an existential threat to us and all other life forms. Fundamentally, by making the body subordinate to what we think means our brain has no direct connection with what we feel in the present moment. Nor do we have awareness of how it feels to exist in this world because the only way we can feel is through our senses. Disconnected from what we feel, our mind functions without a moral compass, making it easy to justify any action, decision, or behavior regardless of its ramifications or consequences.

Evolution dictates that body language is still the primary communication method our brain processes when we interact with one another, whether we are aware of it or not.[5] The body exists in the present and cannot make up a story or lie to conceal what it feels, which means truth is always evident in our "body language". In specialist police interrogations officers are trained to focus on movements such as an eye flicker or facial tic rather than what the detainee is saying. Poker players wear sunglasses because their eyes always give them away to those who are tuned into body language.

Anna Breytenbach[6] has used her amazing ability to communicate through body language to save the lives of wild and domestic animals that have been rescued from abuse and lack of understanding.

In the last decade science, with the advent of MRI technology, has turned on its head the computer mind over machine body belief. Neuroscience, in studying how the brain works, now knows it is a complex neurological organism that rewires itself as it learns and refines skills. Two-thirds of the brain evolved to process all information it receives through the body—millions of bits of data in any given moment. The nervous system has tendrils throughout the body to detect and convey information to the brain from the senses: what we hear, see, smell, taste, and touch.[7] *Every sense is activated by movement.* And since every sense is activated by movement, *movement is the source of all learning.* Neuroscientist and roboticist Daniel Wolpert Ph.D., based on evidence from other animals, goes so far as to state, "The only reason we have a brain is because we move."[8]

One-third of the brain—the ability to consciously reason—evolved later. *It receives all information from the other two-thirds of the brain.*[9] The amount of data it receives in any given moment is minuscule—about 16 - 40 bits.[10]

An excerpt from Edward Slingerland's book Trying Not to Try[11] defines the two parts of the brain this way:

> *So although talk of "mind" and "body" is technically inaccurate, it does capture an important functional difference between two systems: a slow, cold, conscious mind and a fast, hot, unconscious set of bodily instincts, hunches, and skills. "We" tend to identify with the cold, slow system because it is the seat of our conscious awareness and our sense of self. Beneath this conscious self, though, is another self—much bigger and more powerful—that we have no direct access to.*

Science doesn't know exactly how the sensory self determines what fraction of information is passed to the conscious self, but through observation and experience it can be inferred that the vehicle it lives in—your body—is a pretty high priority. The nature of the information your consciousness receives depends on how well you move. Science has found that poorly organized movements can regress intelligence while more refined and complex movements increase intelligence. Neuroscientist Michael Merzenich Ph. D conducted studies in the nineteen-eighties that showed the correlation between movement and thinking.[12] In one experiment he showed that as a skill is learned fewer irrelevant parts of the body are used, and that as the movements become more precise and refined, neurons fire faster. He further showed that the faster neurons fire, the faster we think. In another important study Merzenich showed that the ability to think more clearly directly correlates with the quality of our movements.[13]

Children with brain injuries or neurological conditions have the best chance to advance cognitively as they develop complexity and refinement in their movements. Those born in normal circumstances but deprived of human touch and the freedom to move can develop severe cognitive limitations. *If children do not move their brain does not develop.* I have seen children whose prognosis was quite dire recover full cognitive skills as they gained movement functionality. Some of the cases are even well documented, but tragically the medical system chooses not to recognize movement quality as an integral part of recovering cognitive and physical functionality. Hopefully this will change soon. In the meantime, through the practices in this book children and adults can become more functionally independent starting right now.

What makes the brain particularly powerful is its ability to take a collection of neurological connections and group them together to form "data packets". The "data packets," (habits) allow us to repeat movement or thought sequences without re-creating them each time. Habits let us do many functions and tasks simultaneously without conscious awareness, leaving the conscious self free to process new information. Habits enable us to play an instrument, walk, drive, cycle, write, and more. Picking up an object, such as a glass of water, seems simple to you and I, but engineers that program robots know otherwise. Picking up a glass of water requires an immense amount of instantaneous processing to determine weight, velocity, trajectory, shape, and more. Today the fastest computers in the world cannot calculate how to pick up different objects in the same amount of time it takes our brain to do it. Computers, much like the brain, need to become self-learning neural networks that make connections and form habits to accomplish complex movement instantaneously. Advances in artificial intelligence are getting closer to emulating the brain by creating neural networks that are self-learning and able to store hierarchies and sequences of function, like habits.[14]

Habits are only as good as your awareness in their creation. The less aware you are, the less functional habits you make. Over time, poor habits can be the underlying cause of pain, limitations, and rigidity—physically, emotionally, and mentally. Injuries, surgeries, adversity, and stress from living in an industrialized world create cellular memories in your body. Time heals the physical injury but your movements remain stuck in the cellular habit resulting from the trauma. Over time you layer on additional poor habits in a cycle that may eventually lead to chronic pain or severe limits in movement. I saw a client that had casts put on his lower legs and feet when he was one-year-old to "correct" a pigeon-toed walk. What it did was freeze his ankle joints and stiffen the feet. Although his body was stiffer than most kids, because the brain is amazingly resourceful, he could do most activities other kids could do. However, over time he experienced more injuries, pain, and discomfort until, in middle-age, he became somewhat disabled with chronic sciatica. To resolve the pain, he has had to get rid of many layers of poor habits and create many new ones to reintegrate his feet and ankles into all his movements.

Injury and trauma are held in the body as sensory memories. Science can't decide whether sensory memories are stored in synapses or brain cells, but new evidence suggests that brain cells do contain cellular memories.[15] Regardless of how, it is clear memories are stored in the body. A tragic accident left a young man with no ability to make long-term memories. He could not consciously remember anything for more than a minute. But he could make and store sensory memories because after his accident he learned new carpentry skills and continued to refine and advance them every day.[16] In my practice, when clients release stuck habits in the body they will often spontaneously remember the trauma or injury associated with the pattern they just released. In other instances of release clients say that they feel like they just got out of jail emotionally and physically. The beauty of resolving trauma through the body is that you are left with just a memory and no associated emotional charge or physical limitation.

To change a habit takes presence and awareness: presence to catch yourself in the habitual act and awareness to feel how to change it. When you change a habit, you weaken old neurological connections and strengthen new ones until the old, less functional connections disappear. There are four ways to identify habits and change them:

- **Connect with how it feels to move**
- **Respond to what you feel the way your brain expects you to respond**
- **Understand how your structure is optimally organized for movement**
- **Integrate five essential elements into all movements**

Up to five essential elements are intrinsic to every action or movement we do. They are the foundation of how we evolved to move in gravity, while adhering to physics and Newton's third law of motion. One or more element—how we breathe, how we yield to surfaces (like a floor) and yield within our structure, how to differentiate movements through the joints creating flexibility and agility, how the structure is round and naturally moves by rotation, and finally, how propulsion moves the structure—applies to any given movement. Though children learn how to move non-consciously using the five elements, adults must consciously understand and feel each element to apply them appropriately to enhance and improve movements.

As you understand and feel how to integrate each element into your movements you can incorporate them into every practice in this book. Since the practices are not exercises you don't have to stop what you are doing to do them. The practices do not address specific skills or activities because they are fundamental to *every* skill and activity. Connecting the mind with the body as the brain evolved to learn, and enhancing skills by integrating the five elements into them will get you in the zone where lies your best potential to excel in any endeavor.

2 - Feel Your Movements

The most important kind of freedom is to be what you really are. If you give up your ability to feel, then in exchange you put on a mask. There can't be any major revolution until there is a personal revolution. It's got to happen inside first.   - Jim Morrison

These days adult teaching methods focus on mechanics, technique and what muscles to engage to learn or refine new skills. I wonder how many of you struggled as I did to learn this way? For instance, I took many hours of horse riding instruction from world class trainers who tried to teach me to be a better rider. I knew my horse wasn't the problem because she moved beautifully when any trainer rode her. Instructors explained the mechanics of movement sequences until I knew them in my sleep. When I rode, they would tell me how and when to engage muscles, but by the time I heard the instructions and tried to apply them it was always too late in the movement sequence. I could not consciously react fast enough to get the timing right. One day in desperation I asked a former Olympian to explain step-by-step a particular movement sequence so I could know ahead of time when and how I should engage muscles within it. She became more agitated as she attempted to explain how to do it until she finally said, "You can't do it like 1, 2, 3, you have to feel how to make adjustments as you do it!" What was I supposed to feel, and when should I feel whatever I am supposed to feel to make the adjustments? At that time I just didn't have enough awareness to feel what was needed to improve.

One thing did become clear; I was unable to improve by focusing on which muscles to engage in a movement sequence. Years later, my intuition that we don't learn by focusing on engaging muscles was confirmed when I worked with children and adults that needed to relearn lost movement functionality. No child will learn how to crawl, walk, roll over, lift their head, or sit up by telling them which muscles to engage. If it were possible to learn by focusing on which muscles to engage then anyone who has had a stroke or brain injury ought to be fully functional again, and everyone would be expert at whatever skill they chose from watching videos or taking instruction.

We are encouraged to feel some things when exercising or learning new skills. I remember being encouraged to "feel the burn" as I pushed myself to improve in a physical skill. I was told sore muscles meant the body is building strength, and strengthening isolated muscles would prevent injury. Clients say they were told that to relieve low back pain they must have strong abdominal muscles to increase "core strength". After surgery or recovering from injury doctors and physical therapists focus on strengthening only the area that was damaged. I believed to get better at a skill required hard work and sometimes pain—the "no pain, no gain" mantra is what I followed. Science and children show us that we learn best when it feels effortless. Learning regresses if we are in pain and discomfort as sore or overused muscles make us weaker[17] and we create poor habits of movement when we ignore pain. Most of us have overdeveloped abdominal muscles caused by sitting and belly breathing. Overdeveloped abdominals prevent critical rotation and counter-rotation in all movements and are usually the underlying cause of low back pain.

In the last sixty years or so stretching has become the accepted technique to becoming more flexible. However, recent studies show that static stretching is harmful to your muscles.[18] Animals and children show that repeating different stretches is not natural. Technically, stretching is the action of pulling on contracted muscles until they warm-up. Once warm they expand, which makes you feel more flexible for a while. After not moving for a while the muscles cool down and you once again feel stiff. I never could stretch as it seemed to always make me even stiffer. I became flexible by learning to *release* muscles within movements. The advantages are that I don't wake up stiff, I don't have to take time out of my day to stretch, and I am almost never sore after heavy exercise.

When we focus on engaging muscles we are unable to focus on the quality of our movements because we cannot feel them. By not feeling movements we tend to transfer the notion of feel to our emotions. Our language and society almost exclusively describes emotions. However, the feel that comes directly from movement is the *source* of emotions. The information our brain is sent through the nervous system from our movements is used to create a sensory map of the world and everything in it. For instance, we have a sensory map of what an apple looks like and how it feels, smells, and tastes. As movements become more refined connections to and between each sense become more complex, resulting in a richer overall experience. In a more complex sensory world we can distinguish between an apple that is crisp or soft, ripe or unripe, sweet or sour, and so on.

The sensory map can become compromised in a variety of ways. When disconnected from feeling our movements, any new skills we learn only partially fill the sensory map. Skills we already have can devolve as our movements become less functional and we lose valuable information in the sensory map. Other ways we can compromise the sensory map is through trauma, injury, and neurological conditions. Restoring the sensory map includes every sense. If we don't include every sense the resulting sensory map will have gaps. A client I will call "Jane" had a traumatic brain injury that wiped out much of her memory and function. Her mother, using picture cards, painstakingly retaught her how to write, speak, and identify objects. Jane also re-learned how to walk through instruction from her mother. Over the years she had become less balanced and started using a walker. She was brought to see me to improve her mobility and function. I watched her walk and gave a couple of instructions. She walked as instructed for a bit and then went back to her habitual walk. I asked her to walk the non-habitual way and then the habitual way and tell me how one felt different from the other. She could not feel any difference! Jane did not know what it *feels* like to walk.

Her caregiver noticed other behavior anomalies. When food shopping she might ask Jane to get some apples. Jane always chose green but she liked to be told which green ones to get. If the caregiver said get the ones that smelled best or felt crispier, Jane brought back the ones she had gotten before, or none at all. To test her sensory map, I asked Jane to close her eyes before I put an apple in her hand. When I asked what it was she could not tell me. With her eyes closed I put a slice of apple in her mouth, and when asked what it was she was eating she did not know. When I showed her a red apple, she wasn't sure what it was. Her sensory map was partially filled only with what she could see, and even that was limited by which picture cards she had been shown.

Unlike the other senses, what you see is not directly connected to the part of the brain that processes what you feel through movements. To populate the sensory map, *what you see must be felt through the other senses.* Scientific studies have shown children with autism or olfactory damage can enhance their sense of smell by improving other senses such as speech and coordination.[19] Clients who have had a stroke express frustration that no matter how long they look at their hand and will it to move, it won't. Sometimes the hand will move a bit by looking at it because some ability to move is in the sensory map, but they have little control of the movement. However, if I ask them to close their eyes they are uniformly shocked to discover they don't feel the hand or arm at all! They will never control movements of the affected hand until information from the other senses populates the sensory map.

Feeling movements is non-conscious for children as they do it naturally, but as adults we must consciously reconnect with what we feel until it eventually becomes non-conscious. The first step toward non-conscious feel starts with consciously becoming aware of and feeling our movements:

- As a baseline, in sitting notice what you feel in your body. Are you slumped or sitting stiffly? How are your legs organized? Are your buttocks relaxed or tight? Where do you breathe from, belly or chest? Is your breath shallow, forced, or normal? Is it easeful to breathe or does it feel constrained? Are you comfortable or do you have twinges or little discomforts? Are your shoulders and neck tight or loose? Is your jaw slack, clenched, or tight?

- Pour a glass of water. Before you pick up the glass think about what muscles you might use to lift it. Once it is clear in your mind, try picking up the glass.

- Did you engage the muscles you planned to use? Did you feel more or less muscles working than you thought you needed?

- Slowly pick up the glass again. How tightly do you hold the glass? Did you stop breathing? Do your shoulders contract or lift? What do you feel in your neck, eyes, jaw, belly, hips, buttocks, and legs? Can you identify what muscles you actually used to lift the glass of water? What muscles do you need?

You may not feel muscles working because the movement is a habit and feeling it has been filtered from your consciousness. Since your consciousness receives just a fraction of what the brain processes, you are limited to feeling just a few things in each moment. To identify which muscles are contracting try the following:

- Touch parts of your body as you lift the glass to feel those muscles that are contracted and those that are released. Touch obvious parts first, then touch parts that don't seem as if they should be involved.

- Wrap your hand around the glass tight enough to feel your muscles. Pick it up keeping the muscles contracted. Does it feel harder and is your arm stiff at the elbow? Are your neck, shoulders, chest, belly, buttocks, or legs contracted? Do you stop breathing?

- Holding the glass firmly raise it up and down a few times and notice what muscles you feel working. Are you breathing to coincide with the raising and lowering of the glass? Do you breathe at all as you raise and lower it? Does your arm feel stiff? Is your shoulder bunched up? Are the belly, buttocks, or legs contracted?

- Does your mind feel more open or closed; darker or lighter; distracted or constricted? Does your body get tighter as you repeat the movement? Do you feel stronger or weaker? Are you less aware of what is around you? Pause.

- Pick up the glass one more time in your normal way and notice if it feels different from the first time you picked it up.

Further contracting and touching muscles are two techniques that help you to *feel* movements. However, there is more to "feel" than becoming aware of contracted muscles because you must also understand *why* they are contracted. Muscles can be contracted due to overuse, but if your structure is not well organized for a movement they can also be contracted due to misuse. To improve movements you must *understand how the structure is best organized to move.* Once you feel a movement you must determine whether it is congruent with the way your structure is best organized to move:

- Lift your arm up as if you are reaching for something. What stops the arm from reaching higher? Do you feel stiff, sore, or tight in the arm itself? Is the neck or shoulder contracted?

- Let the arm go for a moment and this time when you reach go slowly and notice which muscles you feel. Further contract a muscle you feel by moving the arm in a way that engages the muscle even more. Where else in your body are you using muscles? Can you feel the belly muscles tightening? What about your other arm, or your buttocks, ribs, or legs? Does contracting muscles further make it easier or harder to reach? Do you hold your breath? Do you lift your arm like the person in Figure 1? Can you think of a way to lift the arm that might be easier?

- Try lifting the arm in other ways and notice what feels easier and what does not.

Figure 1. Arm reach

The way the shoulder muscles are engaged in Figure 1 limits the usefulness of the arm. Are the shoulder muscles bunched up because of habit, or are they bunched up because of the way she lifts the arm? The way an arm is lifted should always be organized for useful actions such as supporting the weight of an object, pushing or pulling, and reaching further. The chance of injury or accident is greatly reduced as the brain easily and instantaneously adjusts accommodate various actions. It must be considered when lifting an arm whether it is best organized to bear the weight of an object. All land animals that have a skeleton are designed to bear weight through the skeleton. Ideal weight bearing occurs when objects are balanced through the center of bones. Pick up a weight and try lifting your arm in such a way that you feel your bones supporting it rather than just muscles working. You will know because your muscles won't feel fatigue, tremble, or get sore. Another way to feel if you bear weight in the center of bones is through walking. Try the following practice to feel how weight travels through your body to walk:

- Stand with your feet hip distance apart. Notice where in your feet you feel most of the weight. Is it on the inside, outside, heel or ball of the foot? Do you feel pressure on the inside, outside or center of your hips? Do you feel any connection to your spine?

- Slowly take a step, and as you place your foot on the ground notice how the pressure travels through the leg.

  - Does it travel up the outside of the leg from the outside of the foot?

  - Once it gets to the hips where do you feel pressure? Is it on the inside, center, or outside of the hip?

  - Do you slightly lean to the side to transfer weight into the leading foot?

  - With the leading leg on the ground notice where your pelvis is. Is it in front of the back foot or over the front foot?

  - Where does the weight go through your foot as you put the leg down?

  - Do you feel any effect on the spine?

  - Do your muscles on the side of your waist contract as you put the leg down?

  - Is your head leaning forward?

  - Do you lock your knees or stiffen the ankles?

If weight is on the outside of your foot it must also be on the outside of your hip. There are no bones above the outside of your hip, so what is supporting your torso? If you are older your waist and torso is probably getting thicker and blockier because muscles above the hip must become weight bearing. Compare the two sets of legs in Figure 2. The person on the right clearly moves with her weight on the outside of the legs, as shown by overdeveloped outer leg muscles and underdeveloped inner leg muscles. Notice how the legs on the left are evenly muscled. Apparently we do admire evenly developed legs, because they belong to famous World War II pin-up girl Betty Grable.

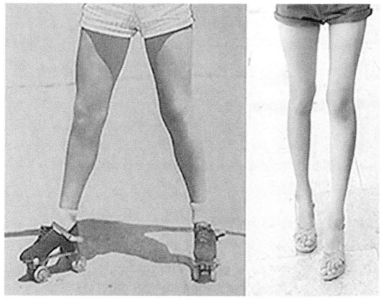

Figure 2. Leg shape comparison

Learning to feel how you move is critical because *you can't change what you don't feel.* To recap the techniques to feel:

- Contract muscles more than is your habit

- Place your hand on muscles that you cannot feel to determine whether they are working

- Try different ways of moving within a sequence to feel which way is more congruent with how your structure is best organized to move

Becoming more aware of what you feel as you move leads to faster and better improvements. The following Movement Awareness guideline helps you identify what you feel to facilitate faster changes. Download and print[20] the guideline so you can refer to it often.

Movement Awareness Guidelines

1. **Establish a baseline for any movement sequence**

    Move in your habitual way and notice what muscles you feel, what your range of motion is, and how much effort it takes to do the movements. From this baseline, you can determine what you need to change.

2. **You can only change what you feel.**

    If you don't feel chronically contracted muscles, you cannot change them. To feel contracted muscles, exaggerate movements by contracting more than usual.

3. **Change within movement**

    It is irrelevant to your brain and to whatever skill you refine whether you release muscles before or after a movement. You must feel what to change within the movement itself.

4. **Close your eyes**

    How it feels to move is not visual, it is sensual. Open eyes distract your brain from sensing and feeling your body move. If they must be open have a soft, diffuse gaze.

5. **Move slowly**

    Slow down to give your consciousness a chance to feel what muscles to release as you repeat movement sequences. Pause within a sticky movement sequence to release muscles before continuing, giving you a chance to smooth it out so the whole sequence becomes more fluid.

6. **Breathe independently of movement**

    Breathe using your diaphragm, separate from your movements.

7. **Be creative**

No practice is rigid or narrowly defined in this method, so feel free to try different variations.

8. **Allow integration**

    The desire to stretch is your brains way of getting back to its habits. Resist the temptation to do any muscle contraction or stretching (which is the same thing) for at least five minutes to allow your brain to integrate the changes.

Using the Movement Awareness Guideline, prepare to lie on the floor for twenty minutes and do a series of non-habitual movements designed to maximize what you feel. You can also download an audio version of the practice.[21] Make it part of your daily routine until the movements become seamless, smooth, and fluid. Repetition of the How to Feel practice integrated with the Movement Awareness guideline will lead to faster, better transformations in all activities.

How to Feel Practice

1. Lie on the floor on a carpet or a mat. Close your eyes and sense what your back feels like touching the floor. Do your heels, shoulders, head, or buttocks press into the floor? Notice how heavy or light your body feels. If your lower back is not touching, notice how far off the floor it is. Where do you breathe? Does your stomach area lift up and down when you breathe, or does your chest? Is your neck curved off the floor? Does it feel long, short, thick, or thin? Is your chin down or up? Does your neck and jaw feel tight or loose? Is your face relaxed or tense? Use your hands to feel under your lower back or anywhere else that you are not sure about.

2. Slowly turn your head left and right. Sense how much effort this takes. Does your head turn with the chin staying on the same trajectory? This time allow your head to fall right and left, letting gravity do the work for you. Did your chin stay in the same trajectory? Notice how much effort goes into the turn and whether your head turned as far as before. As you slowly roll the head

from side to side, try to reduce muscular effort. Notice if you stop breathing as you turn. Try to breathe in and out as you turn the head but don't "time" your breathing to the head turning. Notice if your jaw contracts. Do you contract the muscles around your eyes? Do you frown in concentration? What can you feel at the base of your neck? Keep reducing muscular effort and notice what changes.

3. With eyes closed, turn your head to the left as your eyes turn to the right at the same speed. Did you stop breathing? Repeat many times and notice what happens with your neck muscles, face muscles, shoulders, back, buttocks, belly, legs and feet. Notice if the movement becomes smoother and easier as you release muscles you can feel. Doing a non-habitual movement such as this challenges your brain to be not so strongly attached to its habits.

4. With eyes closed, turn your head to the left as your eyes turn to the right at the same speed. Just out of center your head should "fall" left with gravity. Feeling how to allow gravity to work for you instead of resisting it gives you greater power and control in movement. Repeat and feel how to release all muscles around you head, neck, shoulders, ribs, belly, and pelvis. You should eventually feel the head fall left with no interference.

5. With eyes closed, turn your head to the left as your eyes turn to the right at the same speed. Allow your head to fall left as you slide your jaw to the right at the same time. Let the jaw come forward so the bottom teeth are level with or in front of the upper teeth. Repeat many times and notice what happens with your breathing, the neck and face muscles, shoulders, back, buttocks, belly, legs and feet.

6. Bring the chin down toward the chest by lengthening the back of the neck. Let the chest sink down and the lower jaw relax forward. Repeat a few times, feeling how to release muscles in the front of your neck, chest and belly. Breathe a time or two in this

position. Did your breathing come from a different place? Put the chin up and breathe, then put it down and breathe. You should feel a big difference in how you breathe.

7. Lengthen the back of your neck to bring the chin down toward your chest allowing your closed eyes to look up at the same time. Let the chest sink down and the lower jaw relax forward. Repeat a few times, feeling how to release muscles in the face, front of the neck, chest, and belly.

8. Feel how to bring the back of the skull toward the spine. Allow the front of the neck to release and lengthen, release the jaw and feel how to allow the chest to lengthen. Make sure you do not arch the back. Release the belly, buttocks, and legs.

9. Feel how to bring the back of the skull toward the spine while looking down through closed eyes at the same time. Allow the front of the neck to release and lengthen. Release the face and jaw feeling how to allow the chest to lengthen. Make sure you do not arch the back. Release the belly, buttocks, and legs.

3 - Responding to what you feel

Any idea, plan, or purpose will enter the mind through movement.

If you feel pain during an activity how do you respond? Do you stop the activity, or ignore the pain and keep going? Does the level of pain determine whether you stop? How many activities have you stopped doing because they cause pain? Do you believe the activity is the cause of your pain? A few lucky people work on the computer all the time and they never feel pain, while others have all kinds of problems. Some people run with no pain while many others have had to stop running. Is the activity really the problem or is it the way you do it? Pain is how your brain lets you know you are potentially harming yourself. Changing how you move is what your brain is signaling you to do, but who among us has any idea what to change? We respond to pain and discomfort in ways we have been told will help, unfortunately most of the familiar approaches are not congruent with the way our brain works. I used to think feeling the burn, being sore, and pushing through pain and discomfort was what I should feel. I now know that if I move correctly I feel nothing. Logically, why would the brain need to inform your consciousness that you are moving well?

A client complained that he woke up every morning with low back pain. I asked him to show me the position he slept in and what he felt when in his sleeping position. He said he felt a "stretch" in his back. I had him shift his position so that he no longer felt the stretch. He called the next day, ecstatic because his back felt great after a night in his new position. Feeling muscles work, discomfort, and pain is how your brain lets you know you should *stop moving in that way*. Try the following practice to see if the movements get easier by releasing muscles you feel:

- Stand and reach up with an arm. Hold the arm up and notice what muscles you feel. If you cannot feel your muscles, contract the arm further as you lift. Once you have located muscles that you feel release them. Repeat the movement many times as you continue to identify and release muscles you feel. Does the movement become easier as you repeat?

- This time reach the arm up and keep it up there. How high can you reach comfortably? Do your neck, shoulder, ribs, and legs contract as you reach higher? Put a weight in your hand and hold it up until you feel fatigue. Notice what muscles feel tired.

If you do not release muscles you feel they remain contracted and increasingly interfere with the quality of your movements. The result of misusing and overusing muscles can eventually straitjacket the skeleton so that it becomes very difficult to move. The most extreme example I saw was with a man who had been diagnosed with "stiff man disease" (I am not kidding!). He could barely lift a leg to walk and after a few steps he became immobilized. He stood tall like a soldier but utterly unable to move. One set of muscles in his torso was pushing his legs down while another set was equally trying to lift his legs. The muscles had stalemated. Most people are more contracted in the front of the torso so that as they age they bend forward. The forward bend of the torso causes the skeleton to be unbalanced and forces the leg muscles to become more contracted. Contracted leg muscles compress joints and make it harder to flex, but not impossible to move. Older clients complain about difficulty climbing stairs, walking up and down hills, and sitting or coming up to stand. An immobilized pelvis is the cause. Eventually the wear and tear on the joints may lead to injury, surgeries, and even full joint replacement. If joints no longer flex the brain is triggered to fuse them together causing arthritis and other calcification maladies. Hip and knee replacements, rotator cuff and shoulder surgeries, and foot problems are all too common today due to stiff and compressed joints caused by contracted muscles. If the brain can trigger joints to fuse it can also reverse fusing, if we learn to release contracted muscles so that joints can move again. If caught in time many surgeries can be avoided as released muscles free joints to function again.

Other unintended consequences of overused or misused muscles can have more severe long-term effects. A critical role of the skeleton is to bear weight. The focus on muscle strength rather than movement quality has led to muscles doing the weight bearing job, which leads to conditions such as osteoporosis and arthritis. Weight training is the standard approach to build bone density but studies have shown it often does not work.[22] Weight lifting can exasperate bone density problems by training the muscles to do even more weight bearing. Muscles that become weight bearing bulk up which eventually makes the torso look blocky and shapeless. Try the following practice to feel whether you bear weight through your skeleton: (If you cannot jump up and down just initiate the beginnings of the movement as if you will jump and follow the instructions below. The movements should be repeated enough to get out of breath.)

Figure 3. Prepare to jump

- Stand with your feet hip distance apart and jump up in the air once. Do you hold your breath? If not, do you breathe from the belly? Do you lift your chin? What part of the foot do you use to jump up? Does your rib cage get tight? Do you contract your legs, buttocks, belly and more? Do you land on your heels? Does it feel spongy and soft or hard and jolting to land?

Figure 4. Jumping aligned and relaxed

- Jump up and down many times until you are out of breath. As you jump notice if the breathing moves to a different place in your body. As you fatigue do you engage less muscles? Can you feel more of your skeleton absorb the impact of the landing? Does your chin still lift? Pause and rest.

- This time tuck the chin back and down slightly so your neck aligns with the spine and jump up and down (Fig. 3). Notice what feels different. Are different muscles working? Are you breathing differently? Can you more easily jump higher? Is the landing softer and more toward the ball of your foot? Again, jump until you are out of breath. Where do you breathe from now? Pause and rest.

- Try jumping with your chin up and then with your chin down. Compare the two jumping styles and notice which one feels as if you have more potential. Which allows you to land more softly and not on your heels? It should feel easier, bouncier, and more powerful to jump through the skeleton versus with muscles.

Using just muscles to jump causes stiffness and stress in the joints and eventually will lead to injury. In a well-organized skeleton weight and pressure travel through the center of the bones from head to toe. Jumping with the chin tucked aligns the spine and stacks the skeleton upon itself making it more balanced and better organized to bear weight. The skeleton is designed to leverage against a surface while tendons are the "rubber bands" that become taut to launch the body up, much like a slingshot. You can feel the difference because the jump feels flat, heavy, and hard using just muscles versus the jump that feels almost effortless and self-perpetuating, like a pogo stick, when you use the skeleton and tendons. You may have started the jumping practice using just muscles, but by the time you are out of breath your muscles are too tired to continuously work, so the skeleton takes over, which is why studies indicate jumping to be an effective way to reverse osteoporosis.[23] *How you jump* is what really matters to avoid injury and optimize functionality. The following practice helps refine jumping from the center of the skeleton:

- Stand with your feet hip distance apart. Jump up and down a couple of times. Is the weight on the outside of your feet? Is it in your heels? When you land do you feel pressure on the outside of your hips? Does it feel balanced to jump? Can you feel any involvement through your spine as you land? Is pressure on the outside of your knees and hips as you land? Does your chin come up as you lift and land? Do you breathe in sharply to jump up? Do you breathe at all? Is your pelvis behind your heels when you jump? Does your torso fold when you land or do you mostly feel it in your quads and knees? Do you lift your shoulders to jump up? Jump until you are out of breath, then pause and rest. Notice what feels fatigued, sore, worked, or stiff.

- Stand with the feet hip distance apart. Slightly rotate the knees toward your center until you feel weight on the inside balls of your feet. Tuck the chin in and back to align the spine. Keep the chin tucked as you fold the torso down. Stay over the front of the feet and prepare to jump (Fig. 4). Drop the shoulders down as you jump. The torso should fold as you land, and straighten as you jump up. Release your toes as you jump up. Can you feel the spine as you jump? Are the knees softer, less like hinges? Can you breathe out as you jump up? Jump until you are out of breath. Pause and rest, then notice what you feel. Do you feel sore, fatigue, or stiffness? If so, slowly jump up and down while you release all muscles that feel stiff or sore.

- Practice jumping your normal way and then the new way and compare the differences. You should feel a distinct sense of greater bounce and ease jumping up through the skeleton versus the hard effort it takes to use just muscles.

Skeletal weight bearing through your center makes you more flexible, stronger, agile, and powerful. In addition, you are more balanced and less likely to injure yourself. To optimize your jumping ability, you must feel how to engage or release muscles at just the right moment. To refine timing in any movement, your musculature must be fully released in between movements. If you initiate a movement using already contracted muscles, your brain must first disengage and then reengage them, which negatively affects your timing. In addition, chronically contracted muscles are deprived of oxygen and blood and become progressively weaker and less responsive. As clients release chronically contracted muscles they invariably comment that they feel stronger and have more stamina in every activity. Timing naturally improves as your skeleton becomes more balanced through its center in movement. To cultivate optimal timing, you must feel what muscles to release within an activity. Try the following practice to refine movement timing:

- Tap your fingers in order on a hard surface starting with the index finger to the pinkie. Notice how fast and evenly you can tap. Does it feel quite coordinated? Reverse the order and notice if the movement becomes slower, less even and less coordinated. Repeat until you feel fatigue somewhere in your body (most likely your wrist, lower arm, shoulders, and neck).

- This time slowly tap your fingers one at a time as you feel what to release in your wrist, elbow, shoulder, neck, jaw, and arm muscles. Repeat the movement slowly for each finger until you feel your entire arm, shoulder, chest, jaw, and neck release.

- Repeat the movement again and breathe from your chest. Suck the belly muscles in and hold while you breathe. Breathing may be more difficult but at least you cannot breathe from the belly. Continue slowly tapping each finger in sequence on the hard surface and release whatever muscles you feel in your arm, shoulder, elbow, wrist, hand, neck, chest, and jaw.

- Reverse the tapping movement starting with the pinkie finger. Notice if how you breathe changes or whether your muscles become more tense. Release as you slowly repeat, paying attention to how you breathe.

- This time start the movement slowly and gradually speed it up without tensing anywhere else in your body or changing how you breathe. As you breathe, contract the belly muscles for a bit and then release them. Do how you breathe change?

- Tap in your habitual order first and then reverse the order. Notice if there is much difference between your habitual and non-habitual way. As you repeat, notice if you fatigue to the same degree as before.

As you release muscles your coordination should improve and the movement will become faster, smoother, and easier. Hand dexterity is one example of how the skeleton evolved to move in incredibly diverse ways. As children we didn't have to know how the body evolved to move because we learned non-consciously, by feel. By the time we are adults most of us have, to varying degrees, disconnected from learning by what we feel, so now we must consciously understand how the skeleton is best organized to move to change and improve our movements. The following practice is an example of understanding how the skeleton is best organized for a movement and an opportunity to feel why:

- Reach up your normal way. Notice what stops the movement, and how your neck, shoulders, chest, jaw, and the arm itself feel. Put a weight in your hand, reach up and notice how long it takes to feel fatigue.

- Reach the arm up and turn the elbow toward your center (Fig. 5). Notice if that relieves the tightness of the muscles in your neck, upper back, and the top of your shoulder. Reach higher and notice what else becomes involved. Put a weight in your hand, reach up and hold it for a bit. Do you fatigue as quickly when you hold the elbow turned toward your center versus your normal way of reaching up?

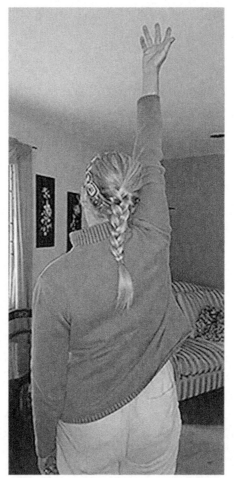

Figure 5. Reach by turning the elbow in

- By turning the elbow inward, you are less able to lock it and more likely to bear skeletal weight. Reach higher and feel how your ribs get further apart as the torso lengthens on one side (Fig. 5). Turning the elbow in and reaching further must engage other parts of the skeleton. Once one independent part hits its limit other parts of the skeleton must engage to further your range of motion.

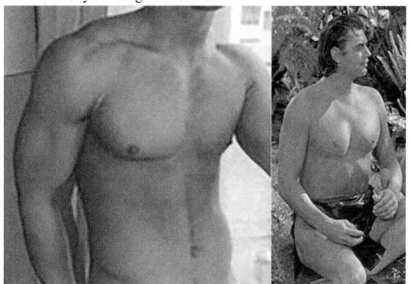

Figure 6. Shoulder comparison

The shoulder is one of the most flexible joints in the entire skeleton, which also makes it more vulnerable to injury from misuse. Only three bones make up the shoulder joint while fifteen plus muscles support it. The tendency today is to build muscles in the arms and shoulders without regard for function. Just as we think the curved spine is "normal," rounded over-muscled shoulders are also a new normal. In Figure 6, notice how the person on the left has bulky, large muscles in back of the shoulder which is forcing the bones forward and down. Range of motion and the ability to bear weight through the bones is compromised. In addition, he can no longer straighten his arms because the muscles are so contracted. What was "normal" fifty years ago is shown by the person on the right, Johnny Weissmuller, an actor and Olympic swimming gold medalist.[24] Notice how upright and back his shoulders are as they rest on the shoulder girdle. He has full range of motion. He can straighten his arms just by seeing how easeful they are resting on his legs. In Figure 6 the bulky muscles of the person on the left is pulling the skeleton forward and down, and eventually will lead to the stooped posture we see in so many adults nowadays.

Gravity is an important force that greatly affects how we function. We work against gravity when we initiate movement by "pushing" the body from behind. To work with gravity, we must feel as if the body is being "pulled" from the front. To get an idea of push and pull in gravity try the following:

- Imagine someone is pulling the arm up by your index finger. As you slowly lift, release your elbow. Instead of pushing the arm up from below using the triceps you should feel it being pulled up using the bicep. Your lower arm, shoulders, and neck have no job in this movement.

- You can have a friend or partner pull your arm up by the index finger. If muscles are released the weight of the arm should feel extremely light. Feel what more you can release when the arm is pulled up by someone else. Then try to repeat the movement yourself and notice if the arm feels lighter.

Our ability to learn and excel derives from the non-conscious, sensory self. However, the conscious self evolved with the capacity to interfere with non-conscious function because we can *choose* to ignore what we feel. I have chosen to ignore pain just to finish running a race, or to get a job done that seemed important at the time. There are times when ignoring pain is necessary, such as a bad injury or life threatening event. Unfortunately, we tend to override or ignore pain so often that it becomes a habit. After sessions clients are often surprised that they feel no pain, so they start moving in exaggerated ways searching for it. I must tell them to pay attention to their comfort zone and stop searching for the pain. It is an insidious cycle that can become our worst enemy as we habitually seek assurance in pain and discomfort. Emotional trauma and other forms of abuse become toxic when we seek negative attention out of habit or learn to ignore any feeling in self-preservation. Therapy and counseling consciously address trauma, but fully resolving any trauma must occur through cellular, non-conscious memory that is held in the body. Neuroscientist Joseph LeDoux and his colleagues, through studies, have shown that the only way we can consciously access our emotions is through sensory feel.[25] I briefly saw a woman who was in therapy for many years to heal from sexual abuse. She was intellectually done, but she felt something was still not right in herself. After a couple of movement sessions, she was elated to finally feel done in her body.

How you respond to what you feel can be summarized as:

- Release muscles you feel
- to establish skeletal weight bearing in gravity
- through the center of your bones
- while releasing joints
- results in better timing of all movements

Changing your response to what you feel will help you transition to function better in all areas of your life. As you transition the mind chatter begins to decrease. When movements become effortless, with full range of motion, maximum power and flexibility, and you feel nothing, there is new space in the mind for other information to enter your consciousness. In uncluttered consciousness you feel and notice a breath of wind, aromas, what you hear, the taste of the air, how you breathe, the quality of touch, and the ease in your movements. Through movement, your body becomes a conduit through which all that you feel from without and within is sensual, alive, and present. Cultivating awareness of what you feel enhances your ability to connect to your own truth and to better connect with others physically, emotionally, intellectually, and spiritually.

4 - Element 1: Breathing

I wake up every day and I think, I'm breathing! It's a good day. - Eve Ensler

Everyone breathes, but it is how we breathe that matters for health, optimal function, and everyday living. Unlike many other animals, we are designed to breathe independently of our movements. Back when we survived in the wild, our ability to catch prey was predicated on the fact that we could run for many hours and miles without exhaustion, whereas our prey could run faster but not as far before they were exhausted. Horses, antelope, and other prey animals breathe through their abdominal cavity, which is controlled by their stride. This means they take a breath with every stride. We outdistanced them because we have a separate breathing muscle called a diaphragm that allows us to take a breath every two, three or more strides.

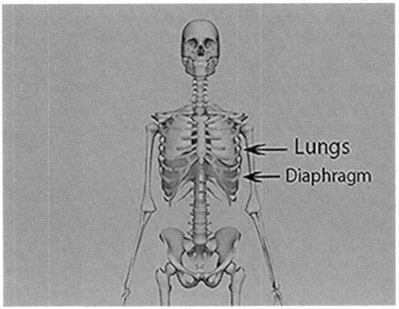

Figure 7. Location of the lungs and diaphragm

To clarify what I mean try lifting your leg up and down a couple of times while you breathe in. Now breathe out as you lift your leg a few times. Change it up to slowly lift a leg as you breathe in and out many times. Then quickly lift your leg up and down several times as you breathe in and out slowly one time. Try walking around and breathe independently of your walk stride. Try breathing in and out once while doing several running strides. Because breathing is so vital to staying alive, control of the diaphragm is in the oldest and most protected part of the body, the brainstem. You can consciously trigger the diaphragm muscle but for the most part it works automatically, such as when you are asleep.

The diaphragm is behind the rib cage (Fig. 7). To inhale, the diaphragm compresses down pulling air into the lungs which lengthens the ribcage, as if a hook is on the top of your head. The pull lengthens the spine, causing the chin to tip slightly down. The ribcage does not lift straight up; it curves out in front and arches at the back in the upper ribcage between the shoulder blades. Notice in the breathing video[26] how the shoulders slide back to make room for air to fill the chest just below the collarbones. Conversely, everything below the rib cage releases down, elongating and flattening the belly. The feeling ought to be as if everything below the ribcage is 'dangling' or hanging down. Stand up and take a breath. Notice if you feel it in your chest or belly? In front of a mirror, turn to stand sideways and watch as you breathe normally. Does your belly protrude out or pull in and lengthen?

How the skeleton is organized to move greatly affects your ability to breathe from the diaphragm. Figure 8 illustrates two quite different MRI photos of spine shapes. The spine on the left is a more familiar shape from pictures and anatomy books, however it shows a spine *held* in its curved position by muscles. The pelvis is tilted forward which compresses the lower spine. Notice how the curvature of the spine causes the belly to protrude. The vertebrae just above the low back are compressed, likely to eventually cause back pain or worse. A tilted pelvis creates imbalance throughout the skeletal structure which means the muscles must be more engaged to keep the individual upright. By engaging the muscles, the diaphragm becomes constricted and breathing is more forced. Forced breathing occurs through the belly muscles below and around the diaphragm.

Contrast it with the spine on the right that is quite straight. The pelvis is horizontally and vertically level, the belly is flat, and each vertebra is evenly spaced. The individual to the right is in balance and no part of the spine is compressed. The diaphragm is unconstrained and free to perform its job.

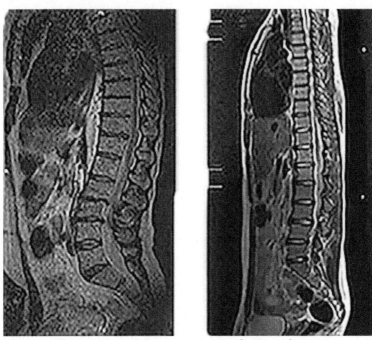

Figure 8. MRI comparison of spine shapes

How is your spine and pelvis organized to stand? The best way to feel it is to lie down on your back with your legs extended and notice what parts of the back of your body touch the floor and what parts do not. If your lower back does not touch the floor you are organized much like the spine on the left of Figure 8. It does not matter whether you stand, sit, or lie down because your brain organizes your body the same habitual way. We know this because, after all, you must lie down to get an MRI.

- Stand up and take a breath. Notice whether the belly moves as you breathe. Breathe deeply once again and notice what feels tight or constrained as you try to draw in air.

- Stand in front of a wall with your toes about 1-2 inches away. Let your body fall into the wall so that your pubic bone, chest and forehead touch (Fig. 46 left). Breathe and notice where the air goes. Do you breathe from your chest (and diaphragm)? Does your chest feel constricted? If so, your ribs likely have little ability to move.

- Next, stand in front of the wall again and this time allow your body to come forward letting just the points of the hips, chest and chin touch. Do you breathe from the belly? Which way of breathing feels more likely to increase your air intake?

In many years of practice, without exception, every adult I see breathes from the belly. I believe we become belly breathers starting quite early because of sitting. Sit down in your normal sitting position and breathe. Does it feel more comfortable to breathe from the belly? If you are slumped or your pelvis is rolled behind you it is easier to breathe from the belly. Sitting on a sofa and leaning back constrains your chest such that belly breathing is your best option. Sometime in the last forty or fifty years' we have decided as a society we are supposed to breathe from the belly. Many clients say that belly breathing is a big part of exercise modalities such as yoga and pilates. Breathing from the belly within your exercise regimen is fine, but the diaphragm is how you evolved to breathe. How do you determine what it feels like to breathe from your diaphragm versus the belly?

- Sit in a chair your usual way and breathe. Notice from where you breathe, the belly or the chest?

- This time tuck the chin in and slightly down and bring the head back until it aligns with the rest of the spine (Fig 9). Do not touch the roof of your mouth with your tongue, let it rest on the lower palette just behind the bottom teeth. Drop your chest but keep the chin tucked and breathe. Where does your breath originate now?

Belly breathing leads to many limitations. For example, most clients do not like to jump because it feels hard to breathe. One reason jumping becomes difficult is due to belly breathing. If you are able, stand up and jump up in the air one time. Did you hold your breath? Belly muscles are needed to jump in the air. You hold your breath because you cannot breathe and jump at the same time using the same set of muscles. If you continue to jump up and down until you are out of breath you will notice the breathing must move to your chest. To maintain oxygen levels your brain must stop the belly muscles from interfering with your breathing. If you are constricted through the ribs breathing may feel difficult, even painful, *yet you must breathe from your diaphragm.* As breathing returns to "normal" you will go back to your habit of breathing from the belly.

Figure 9. Sitting with chin tucked and head aligned

You might think using the belly muscles to breathe allows you to take a deeper breath, however the reality is that it constrains the diaphragm causing you to take shallower breaths. Shallow breathing can irritate conditions such as high blood pressure, stress, stiffness, headaches, low back pain and more. Over time many people lose height as chronically contracted belly muscles pull the rib cage forward and down. Stand up and put your hand between your hip and first floating rib. If it is not at least four fingers wide you are losing height. As breathing becomes progressively shallow, labored, and uncomfortable we tend to exert ourselves less and less. We experience physical limitations including shortness of breath, an inability to jump well, difficulty running or walking very far or fast, and a lack of stamina. Since limitations are usually gradual, we make excuses, such as blaming getting older. Clients with big bellies always tell me it is just fat, but as they learn to breathe from the diaphragm they are amazed to see their big belly get smaller even though they haven't lost any weight. Belly breathing affects how we speak as well:

- Say a few words and notice how you breathe. Do you hold your breath? Do you contract the belly? Do you feel your voice in the back of the throat or in your chest?

- As you speak, contract the belly on purpose and notice what happens to your voice. Does it get more constricted or go into a higher register? Contracting the belly, chest, neck, shoulders and more inhibits movement and air flow which can strain the vocal chords.

- Practice humming as you exhale and release any muscles you feel. Eventually start to speak as you exhale, releasing the belly, chest, neck and jaw, and notice how different it feels to talk. Does your voice feel different? Is it slightly lower, more easeful, less forced? Does it feel pleasant to speak without so much effort? Do you notice you can't get many words out before you need to inhale?

As you refine diaphragm breathing you will be able to breathe more deeply and speak without having to take a breath so often. As you transition, you may notice that your voice lowers into a deeper register, and you feel a rumbling in your rib cage and chest.

Figure 10. Sitting slumped leads to belly breathing

The way we form words using our mouth, jaw, and neck muscles is critical to the quality of how we speak. As you say the sentence--*he feared his vehicle would veer off the road*--notice whether your neck, lips, and jaw feel tight or loose. Does your mouth move much? Does your lower jaw release forward or do you draw it back as you form the words? Does your neck feel tight? Is your posture more like the individual in Figure 10 or the individual in Figure 9? To project the voice out we must be aligned in the spine and learn to use our mouth and jaw to speak forward. Try the following voice projection practice:

- Sit with an aligned spine (Fig. 9), purse your lips forward and say the word "five". Do it several times, releasing your jaw, neck and facial muscles. Allow your lower jaw to come forward as you pronounce the "v" sound of five. Exaggerate moving the jaw as the lower teeth slide in front of the upper teeth. Notice the timbre of your voice as you go back and forth between speaking "five" your habitual way and saying it with your lips pursed and the lower jaw forward.

- As you sit in your usual position, try saying the word "five" several times. Adjust your position to be like Figure 9, purse your lips and repeat the word several times as you feel how to release the lower jaw forward and down. Repeat the word "five" from both positions so you can notice the difference with the chin tucked down versus the chin up (Fig. 10).

- With your chin tucked as in Figure 9 repeat the phrase *he feared his vehicle would veer off the road,* making sure to purse your lips forward as you speak. Feel how to release the neck, jaw, and facial muscles as you exhale and speak. Release the belly muscles, sphincter, shoulders, back, and chest as you speak. Does anything change for you?

The quality of your voice is dependent on many factors: how you breathe, the position of your head, how you pronounce words using your lips, jaw, and neck, and how released the belly muscles are. We like listening to a melodic, deeper, relaxed, rounder voice, which is why many performers and presenters are trained to speak in a lower register. A client who sings soprano struggled to hold longer notes and project certain lower registers. Her voice felt strained after singing for an hour or so. As she transitioned to breathe using her diaphragm correctly, she could hold notes longer, could project the lower registers better, and she could sing for two to three hours without feeling vocal strain or stress. You can improve the resonance, timbre, and tone of your voice just by learning to breathe from your diaphragm and speak in a forward projection from your lower jaw.

A common mistake people make when transitioning to breathe from the diaphragm is to use the upper belly muscles to lift the ribcage up. Lifting the ribcage with muscles pushes your chest in front of the vertical which causes the pelvis to tilt forward. Try lifting your ribcage up and notice what happens. Does your weight shift into the heels of your feet? Breathe with your chin tucked, the back of the neck elongated, and belly muscles released, and then breathe by using muscles to lift the ribcage. Notice how your balance and breathing are affected.

Many clients say that as they breathe with more ease, their mood becomes lighter and they feel less stiff or constrained in their body. Some have relieved or eliminated conditions such as high blood pressure, angina, asthma, chronic fatigue, tinnitus, sleep disorders, hypertension, acid reflux, digestive disorders, bowel irregularities, migraines, headaches, and hypertension. Others have noticed that their mind is clearer and they think better. Performers notice they are less nervous or anxious and their timing is better. Get back to breathing the way you evolved, by using the diaphragm muscle, and transform your life in every way.

5 - Element 2: Yield

Optimize connection, timing, power, and strength through yield.

Yielding through the body facilitates optimal mobility, timing, strength, power, and agility. Yield adheres to Isaac Newton's third law of motion[27] which says: *When one body exerts force on a second body, the second body simultaneously exerts an equal force in the opposite direction.* Energy enters the body from one direction and, depending on the intention, it can be absorbed to stop a movement or increased to make a movement bigger. Jumping best illustrates energy increase and decrease:

- Jump up and down with the intention of jumping higher each time. Be sure to jump keeping the chin tucked and head aligned on the spine. Do you feel heavy or jarring on landing? Do your knees lock when you jump up or land? Notice what muscles you feel and try to release them each time you jump. Does releasing muscles change the feeling of heaviness? Notice if releasing muscles faster improves timing and height. Pause.

- This time jump up with the intention of landing and not jumping any more. Do you bend your knees more or less? Does it feel heavy or jarring to stop jumping? Do your knees feel stiff when you land? Try the jump a few more times and release muscles you feel. Notice if the landing becomes softer and spongier on landing. Does the landing get softer if you let your knees slightly roll toward one another?

The first practice increases energy for the next jump while the second practice absorbs and decreases energy to stop jumping. If your body is unnecessarily contracted you lose energy when jumping higher or you block energy and prevent a soft landing. Blocked energy is what makes jumping hard, uncomfortable, or even painful, which is why most adults do not jump. By learning to yield through the body jumping can become easy and powerful. To integrate yielding into every movement three different modes must be considered:

- External yield – Leverage against a surface to move the body

- Yield from within – Leverage within the body to move other parts of the body

- Structural connection – How we skeletally touch and connect with one another.

External Yield

Figure 11. Runner leverages a surface (block)

You can leverage against a surface to move the body away from it as shown by the runner in Figure 11. Forward momentum should be equal to the backward push. If muscles are incorrectly contracted momentum is blocked or diminished. Try the following movements to feel how to leverage against a surface:

- Sit on a bench or the floor. Bend to your left side and place your elbow down by your side and lean on it. Begin to push the elbow into the surface. Does your torso move toward the elbow? Do your ribs bend in any direction? Is your neck, ribs, shoulders, belly, inside of your thighs contracted? Is the shoulder jamming into your neck? Does one sit bone come off the table or floor? Do you push through your feet? Pause.

Figure 12. Pushing the elbow into a surface

- This time push the elbow into the table or floor with the intention that your torso must move away from it. Release any muscles you feel as you repeat the movement. Notice as you release the parts of the skeleton that move. Do the ribs bend? Does your left shoulder drop down toward the table or floor? Does your head drop down toward the shoulder? Does your pelvis move toward both sit bones touching the floor or table? Do you feel the more you release muscles the easier the skeleton moves? Do you feel more weight bearing through the arm?

In Figure 12 the individual is pushing her elbow into the table with her body. Notice how the ribs angle down and the pelvis tips toward its side. In Figure 12a she is using the table surface to leverage her body away from the elbow. Her ribs bend, the pelvis rests on the table and her shoulder is lowered allowing her neck and head to drop down. You can only feel how to leverage one part of the skeleton to move another part if you release muscles such that the skeleton bears weight. Power and strength increase along with better timing, agility, and mobility. A runner whose muscles are optimally released in the starting block is the readiest for explosive propulsion off the block.

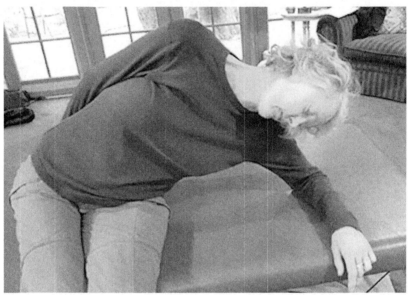

Figure 12a. Leverage the elbow to move another part of the body

If muscles engage incorrectly within the skeleton movements such as walking can become more difficult. Common complaints I hear is that legs feel heavy to lift, especially going up steps. Another common complaint is tight hamstrings, pain on the outside of lower legs, and painful feet. The cause of heavy and painful legs is contracted muscles in the torso pushing the legs down while trying to lift them. Try lifting a leg as if you are going to take a walk step. Does your opposite leg stiffen and brace on the floor when you lift? Do you contract the belly muscles? Does your chest and neck stiffen? Lift your leg up and down a few times. Does the leg feel light or heavy? Bracing and stiffening as you transfer weight from one side to the other affects your balance so that you tire sooner and become more susceptible to stumbles, trips, shuffling, and falls. Try the following movements while imagining a different perspective:

- Imagine the floor beneath is pushing up against your feet such that you must release your torso down as the floor rises. The bottom of your feet soften by releasing your ankles, knees, hips, belly muscles, ribs and neck. Lift one leg and then the other as you imagine the floor pushing up. Notice if lifting a leg feels easier or lighter.

- Now imagine you are standing on a small boat with waves making it bob around. Feel how to stay balanced by releasing through your ankle, knee, and hip joints as the boat undulates beneath your feet. Lift a leg every so often and notice if it feels lighter and more flexible.

- This time imagine you are walking on a carpet suspended a few inches off the floor. As you put your foot down you must feel how to not push against the carpet so it doesn't sink down. Walk a few steps and notice what you need to release and soften to stay suspended.

A non-solid floor beneath your feet compels you to release throughout the body to accommodate the softness. You must release muscles or you fall over. The practice highlights how to yield to a soft or hard surface. Releasing muscles allows your skeleton to bear weight again which facilitates adhering to Newton's third law of motion in movement.

Yield from Within

One part of your body can leverage to move other parts of the body. The application of Newton's law of motion is a bit different here because we are only working within the body. The rules are the same, one part equally moves another part, but within the body itself. The following movement sequence should clarify what I mean:

- Drop to the floor on your hands and knees. Sit your buttocks back on your heels and bring your elbows in front of your knees. Begin pressing or leveraging the elbows into your knees to lift the pelvis up and forward (Fig. 13).

- Is your spine flexing?
- Are you using belly or leg muscles?
- Do your shoulders contract and hunch forward?
- Do you hold your breath?
- Is your neck contracted?
- Do your feet come off the floor?

Figure 13. Using muscles to lift the pelvis

- This time try to release every muscle you feel as you imagine the pelvis is being pulled up by your tailbone. Your spine should feel like it is lengthening and flattening down toward the floor. Make sure your lower legs are released so that your feet stay on the floor (Fig. 13a). Release your chest, belly, shoulders, neck, and legs. Have a partner help by pushing your sit bones up and forward (Fig. 14).

When you use muscles to lift the pelvis your back flexes, the feet come off the floor and the neck strains (Figure 13). Contrast it with allowing your arm bones to leverage against the knees to lift the pelvis so that the spine is straight, the feet and neck relaxed, and the movement feels effortless (Figure 13a).

Most movements we do rely on a measure of independence of the torso above the pelvis from the legs below. For instance, in walking your legs and pelvis move while your torso stays facing in the direction you are moving. The torso must be loose and released so that it can counter-balance the movements of the pelvis and legs. To get a sense of the upper and lower half of your body as independent, try the following:

Figure 13a. Leveraging the skeleton to lift the pelvis

- Stand with your feet hip distance apart. Imagine a hook is attached to the top of your head and it is pulled just enough to almost suspend you from the floor. Breathe from your diaphragm and notice if your body below the ribcage can drop down as the ribcage lifts to suspend your body. Your lower half feels like it is dangling down while your upper half is suspended and lifted by the action of breathing through the diaphragm.

- The hook slowly lowers you so that you must feel how to allow your bones to support you in standing with minimal muscle support. Continue the diaphragm breathing as you feel whether you resist and push against the floor by stiffening muscles around your joints. Practice many times breathing and feeling how to release muscles and allow the floor and your bones to support you.

- Once you feel it try taking a walk step. Do you stiffen and brace the standing leg against the floor as you lift the other leg and put it down? Stiffening the opposite leg compromises your balance. Feel how to release both legs as you repeat the movement many times.

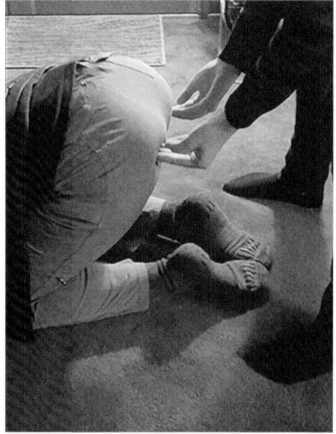

Figure 14. Push sit bones to lift pelvis

Releasing muscles as you leverage the skeleton within itself insures you will not block energy in any movement. You get a better feel for weight bearing through the skeleton as well as optimizing functionality regardless of the activity.

Structural Connection

A handshake or a hug tell us a lot about one another. Most of us have experienced the bone crusher versus the wet dishrag handshake or hug. Are you aware of what a quality hug or handshake feels like? Perhaps you remember someone whose handshake or hug left you with a favorable impression. How we touch is a powerful tool of communication that, if we are aware, contains a trove of information about one another. Whether you know it or not your sensory self responds best to a structural connection. I felt a structural connection for the first time in a movement training when one of the instructors shook my hand. I was compelled to move wherever she directed me, not in a forceful way, but like a guide. My muscles released such that we became two people moving as one. The "art" of all martial arts is to learn a structural connection such that you are in an equal dance with your partner. The T'ai Chi Ch'uan practice of pushing hands (Fig. 15) is a great way to understand its power. Adhering to Newton's laws of motion, the goal is to place your hands in front of you, palm forward to your partner, and apply pressure equal to the pressure they apply to you so that you cannot tell who is leading and who is following. The hands move in all directions in space as one if you are equally balanced and yielding to one another. Should one person have an intention to move quickly, the other can potentially feel the intention before it happens, and be ready to respond. It is an awareness that is present in every moment in your life and is fundamental to excelling in all that you do.

A structural connection is non-conscious because your conscious self is too slow to respond to the touch of another. Experiments by psychologist Benjamin Libet and others measured a small delay before the conscious self receives data from the sensory self. In one experiment they wired subjects with measuring electrodes and then asked them to move a finger.[28] A half-second before the decision was made to flex the muscle, an electrical signal was detected in the brain. The "decision" seemed to be made by non-conscious neurons before the conscious self became aware of its desire to act. Libet also showed that it takes a full half-second for us to consciously become aware of an event such as a finger pricked by a pin. Another experiment showed subjects pulling their hand away from a hot burner before they consciously registered that it was hot. A half second does not sound like much time, but in the context of actions such as jumping out of the way of a vehicle, or hitting a ball traveling over 110 mph, it is a long time indeed. It also indicates that most of our actions cannot rely on the conscious self.

Figure 15. Pushing hands in martial arts

How we touch is what makes us sensual beings as we yield from within to increase awareness of what we feel from without. When you are sensually aware every millimeter of your body has a heightened sensitivity to what you feel externally and from within your body. How we touch affects the quality of how we connect emotionally and intellectually with one another. Try the following to see if you can feel a structural connection using a tennis ball:

- Sit on a chair with a tennis ball in your hand. Lean forward and rest the elbow of the arm holding the tennis ball on your thigh. Gently begin to squeeze the ball in an increasingly snug hold. Notice if your wrist, arm, elbow, shoulders, neck, or chest contract? Do you hold your breath? Does your hand and lower arm look similar to Figure 16?

- Try again and this time feel how to wrap your hand around the ball without stiffening your wrist, elbow, shoulders, ribs or neck. Is the palm of your hand and fingers stiff? What muscles do you feel engaged in your lower and upper arm?

Figure 16. Scrunched hand ball hold

- Do the movement again as you imagine your palm and fingers are as soft as a sponge wrapping around the ball. Release through your joints and breathe independently of your movements. Release your lower arm and shoulder feeling just the biceps contract as you squeeze the ball. Imagine you are holding a baby or a small animal such that you squeeze just enough to keep it snug in your hand.

- This time use the tips of your fingers to squeeze the ball while allowing the rest of your hand to release and wrap around the ball such that every part of it is touching the ball with equal pressure. Your hand forms itself around the ball more snugly as you wrap ever more tightly around it, your flesh releases and softens to allow bones to hold the ball. Keep an equal bone pressure around the ball.

Figure 16a. Smooth yet firm and snug ball hold

- Imagine your hand bones are slowly squeezing the ball, much like a boa constrictor wrapping around its prey. Release joints such that the hand pressure wraps around the ball evenly. Notice if your biceps are engaged. Go slowly so that you can feel when you start getting stiff in the joints or you contract too many muscles. The squeeze should slowly become more snug without harshness. Notice if your palm and inside lower arm is smooth (Figure 16a).

Many contemporary conditions such as carpal tunnel, frozen shoulder, trigger fingers, and more are the result of muscle misuse and overuse that constrain joints. Hand therapists are very busy these days with young people getting all manner of hand and arm pain from texting. The tennis ball practice puts you on a path to relieve symptoms and eventually eliminate problems resulting from typing, texting, and poor arm use in general. Working with a tennis ball is a great technique to learn to feel a structural connection. To sensually feel a structural connection, you can practice with your own body. Try the following practice to feel a structural connection within your own body:

- Sit comfortably in a chair and wrap your right hand around your left lower arm.

  - Do you contract or become tense in the left arm?
  - Do your neck, chest, or shoulders contract?
  - How tight or contracted does the right arm feel?
  - Do you stop breathing?
  - Do you feel some parts of the hand that is holding the arm more than others?
  - Does the hand feel hard or raspy?

- This time, wrap the hand around the arm using the ball technique you practiced earlier. Notice as you practice whether your left arm is beginning to relax. Are you relaxing the right arm as you wrap the hand around it? Practice until you feel a firm, even grasp of the left arm without being tense anywhere else in your body.

- Start to move the left arm using the right hand.

  - Did your left arm immediately tense up?
  - Did you contract and tense your right side to move the arm? Feel how to move the left arm without tensing the right side or changing your handhold.
  - Does the hand hold feel mushy and is it slipping? Moving flesh and muscles around is irrelevant to

movement so feel how to move the bones of the left arm. The bone connection should feel soft and comfortable. Once you have a bone connection move the arm.

- Is anything different? Is the arm more relaxed?

Once you feel the connection through bones, it dramatically changes how your brain responds. You can use a structural connection to practice handshakes and hugs, and chances are people will respond more positively. The quality of how you touch can be the difference between a so-so meeting and a wildly positive, exuberant, and heartfelt interaction. How you connect with others depends on who and what you are touching, your intention, and the response. Most animals are way more present and aware than humans and they respond well to a light structural connection. Most human interactions require a firm and secure touch, without force, yet snug. You know it when you feel it.

Rock climbers demonstrate optimal structural connection when they hold onto the tiniest of indentations in a rock using just the last digit of their fingers as leverage. Each joint must let energy through such that the strongest muscles are available to support the body. The same is true for playing a musical instrument. A classical violin and flute player was struggling with carpal tunnel, neck and shoulder pain. I showed her how to play with a soft, fluid wrist, elbow, and shoulder joint so she could feel how little muscular effort was needed to play. As she learned to yield throughout her body, pain and discomfort was replaced by a newfound joy and passion for music again. The quality of a structural connection can be the difference between an accomplished musician and an extraordinary one.

Learning to yield in movement leads to better balance and effortlessly transforms any activity to be light, easy, flexible, and powerful. You become synchronous fluidity from within and without. Even if this feels foreign to you now, you knew it as a baby because you could not have learned how to roll over, sit, crawl, stand, and walk without yielding. The ultimate result of yielding is that you don't have to control anything because you are in a dance with all life, where no one living thing takes precedence over another. In yield there is no winner or loser, everyone and everything is an equal partner in the dance, and all interactions are an opportunity to refine and enhance connections.

6 - Element 3: Differentiation

Free joints allow the body to move with optimal mobility and flexibility.

Bones (components) working through joints create a differentiated, flexible, balanced structure. Flexibility, strength, and balance result when components on each side of joints move differentiated from one another. The type of joint determines the scope of differentiation between components. The shoulder is the most flexible joint in the body with just three bones: the scapula and the collar bone form a "joint" to cradle the arm bone. The collarbone is the sole attachment to the ribcage. It is a delicate joint made robust by the support of muscles and ligaments. A joint of such complexity is vulnerable to misuse if differentiation between the arm and shoulder becomes limited. Try the following to see how freely your arm moves in the shoulder joint:

Figure 17. Lift the arm from the elbow

- Lift the elbow up toward the ceiling while letting the lower arm drop down toward the floor (Fig. 17).

    - What stops the movement?
    - Do you hold your breath as you lift?
    - Does your neck feel stiff?
    - Does your head lean or turn away from the arm as you lift?
    - Does the top of the shoulder feel stiff or sore? Does it push into your neck?

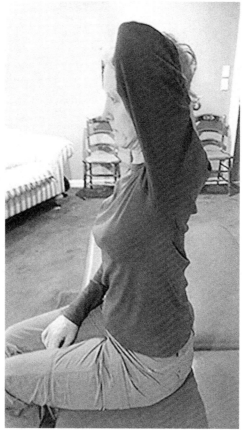

Figure 18. Lift elbow to the ceiling

Your arm may feel constrained in the movement, its use limited, and it can potentially lead to injuries such as dislocation, rotator cuff injury, frozen shoulder and more.

- Lift the elbow and hold. Can you feel what to release in your neck, back, and chest. As you lift the elbow let the top of your shoulder drop down and release. Touch different parts of the body to feel what is contracted so that you learn what needs to release. Repeat the movement many times and release muscles that you feel.

Figure 19. Take the elbow up and down

- This time rotate the elbow toward your center and drop the hand toward the top of the shoulder. Lift the elbow toward the ceiling (Fig. 18). As you lift the elbow, let the top of the shoulder drop down and release. Is your lower arm involved? Keep releasing the shoulder and the lower arm as you lift the elbow and release it down. Rest with the arm long for a moment.

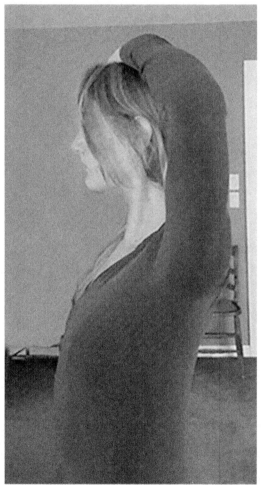

Figure 20. Take the elbow forward and back

- Rotate the elbow toward your center, bend the lower arm so the palm faces the elbow in a right angle. Imagine a string is attached to your index finger pulling the hand up to place the palm on the top of your head. Repeat several times feeling how to release the muscles in the shoulder as the hand moves toward the top of your head.

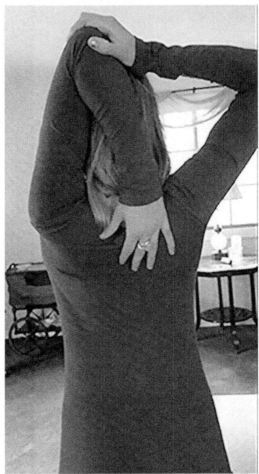

Figure 21. Take the elbow across the body in back

- Repeat the same sequence as before only this time leave the hand on top of your head. Take the elbow out toward your side and begin to bring the elbow up toward the ceiling leaving the hand in place on your head (Fig. 19). No hand sliding allowed! Notice as the elbow comes up toward the ceiling what your head and ribs must do. Next take the elbow down toward the hip and notice what must bend in your torso to allow the movement. Keep the hand in place as you bend the ribs and head.

- Is your neck softening as you bend?
- Does the arm feel tight as you bend side to side?
- Do you hold your breath as you bend?
- What is your opposite side doing as you bend?
- What muscles in the arm feel fatigued as you leave the hand on your head?

- This time with your palm on the top of your head turn the elbow toward the spine (Fig. 20) behind the head. Again, no hand sliding allowed while you notice what gets tight and stiff in your torso and ribs. Leave the shoulders dropped down and make sure you do not tighten or get heavy with the hand on your head. Next turn the elbow toward your opposite shoulder in front of the head. Notice how far you turn in either direction before you can go no further. What is stopping you from turning any further? Keep the hand in place as you turn.

- This time put the hand on your head and slowly slide it down the back side of your head, allowing the elbow to come up toward the ceiling. Use your other hand to cradle the elbow and take it toward the ceiling as you feel what to release (Fig. 21).

- What muscles do you use this time?
- Is your neck and the top of your shoulder involved?
- Do you use your biceps at all?

- What role do your ribs play?
- Repeat several times, feeling what more you can release.

— Try it again, purposely dropping your shoulder down as you slide the hand across your head and down the back side. Does it feel any easier and do you use less of your neck and the top of your shoulder? Once the elbow is pointed toward the ceiling, you can move the lower arm in any direction you like, including bringing it up toward the ceiling. You can, in addition, twist the wrist in any direction as if to grab an object. Notice what feels different lifting your arm in separate components versus as a single unit.

Figure 22. Babies easily bring their legs up to the chest

If you lock joints and try to lift an arm or a leg it feels heavy and stiff. Releasing joints to lift is much easier, lighter and more flexible. By bringing the elbow in towards the torso to reach and push or pull, power passes through the structure while arm strength is maximized using your biceps and torso. Lifting the arm as if it is being pulled rather than pushed from behind reduces your effort. When you push from behind into movement, you must push against your body weight and gravity. When you move as if being pulled, your body moves with gravity, allowing the brain to adjust the structure so that you balance from the center. In centered balanced you are at ease, relaxed, poised, and ready to move.

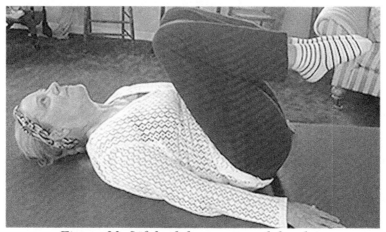

Figure 23. Lift both knees toward the chest

The first preparatory movement babies learn before rolling onto their front is to lift the knees toward the chest (Fig. 22). The ability to bend the legs toward the chest without engaging the upper torso is fundamental to the most common activities we do, such as running and walking. How easily do your legs roll up to the chest?

- Lie on your back and *slowly* lift one leg to the chest.

    - Does the back of your head press into the floor?

    - Does your chin come up?

    - Do you arch the lower to mid-back?

    - Does the back of the pelvis tilt into the floor?

    - What part of the belly works the hardest: upper, mid, or lower?

- If you can bring one leg up without using the upper torso, try bringing both legs up to the chest at the same time (Fig. 23). Both legs at once is quite a bit harder if your muscles are in conflict through the torso.

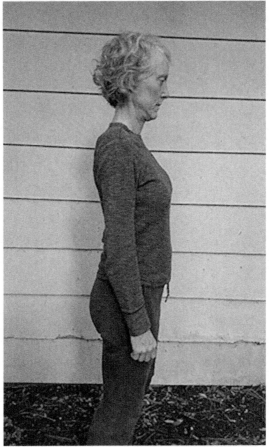

Figure 24. Arched low back, pelvis tilted forward

- Does the back of your head press into the floor?
- Does your chin, neck, and chest come up?
- Do you arch the lower to mid-back?
- Does the pelvis tilt forward toward the floor?

- What part of the belly works the hardest: upper, mid, or lower?

− If you arch the back and lift the chest to bring the legs up, you do the same movement when you walk or run (Fig. 24).

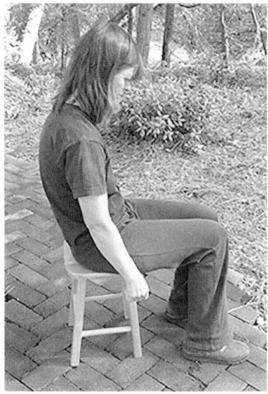

Figure 25. Slump as your torso goes back and forth

Why do we unlearn how to do such a fundamental movement? Sitting may be the culprit. We evolved to squat, but for the last five hundred years sitting in chairs has replaced squatting. When we squat weight bearing is through the feet and the pelvis is the balance point between the upper and lower half of the body. In sitting, balance is through the torso as the sit bones become the new "legs". Your actual legs have no job in sitting, but because the brain organizes you as if you are squatting, all the weight goes through your feet. You must consciously override the non-conscious action of the brain and transfer weight bearing from the legs to your sit-bones and torso. If you sit allowing weight bearing in your feet, the belly muscles become trained to push the knees down toward the floor. The muscles pushing down can eventually become as strong as the muscles pulling the knees up which can lead to a stalemate in the legs and make them impossible to lift. It is not practical to stop sitting, so we must organize to sit balanced without using the legs. Let's start with feeling how to balance through the torso:

Figure 26. Lean forward lifting outside toes

- Sit on a bench with your feet hip distance apart. Reach forward. Do you engage your leg muscles and push through the feet? Reach left and right. Do you push through the feet as you reach? Come forward as if you want to stand up. Do you engage your leg muscles and push through the feet? Does the pelvis have a role in any of the movements?

- Tuck your chin back and down until your neck is even with your spine. Drop your chest without letting the chin come up. Allow your shoulders to round and your pelvis to roll back. Round the torso and begin to come forward without lifting the chin. Go back and forth and do not allow the chin to come up or drop at any time (Fig. 25). The torso should stay rounded throughout the movement. Release your back muscles, the front of your thighs, your buttocks, belly, and chest as you repeat the movement. Release muscles until the movement feels quite effortless. By rounding your torso and keeping the chin tucked, the pelvis must initiate the movement, and it should feel like a ball rolling forward and back.

- This time allow the spine to straighten as you go forward, keeping the chin tucked but *without lifting the chest.* You should feel the back of the neck lengthen and the chin drop down and in slightly, as if you are being pulled forward by the back of the head. Do not arch the lower back or contract belly muscles. Do you push through the feet as you go forward? Try lifting your outside toes up and breathe out as you go forward (Fig. 26). Notice if your shoulders and arms are released or contracted. Release the shoulders and arms as you go back and forth.

- Reduce your range of motion in the movement until you feel a position where the pelvis is balanced, the torso supported and soft, and the legs released. Do you have a sense of weight bearing through your bones as opposed to just feeling the muscles working?

You can use the practice to do actions such as reaching for a phone, a pen, a book, or a piece of paper--whatever you reach for most often at your desk while sitting. Just feeling how to *not push through your feet* will reduce the stiffness, low back pain and other effects of sitting. The *position* of your legs at right angles and hip distance apart is important because they stop you from pitching onto your face when moving your torso around. You may not be aware of pushing through your feet in sitting so the following practice helps you consciously become aware of it and shows you how to stop it:

Figure 27. Lift a leg slumped

- On a flat seated chair, sit forward with your feet at a right angle to your knees and hip distance apart. Lean back into the chair allowing your chin and chest to sink down. Lift one leg and then lift the other (Fig. 27).

    • Do you hold your breath?

    • Does your opposite foot push into the floor?

    • Does your belly come forward or your chin lift?

- Do you initiate by lifting your toes and flexing your foot?
- Does your knee turn out more as you lift the leg?
- How hard does it feel to lift your leg?

Figure 28. Lift both legs slumped

- This time try pointing the foot as you lift each leg. Make sure your chest and head stay down as you lift a leg. You can suck the lower belly muscles in as you lift to make sure the chest and head stay sunk forward.

  - What changed? Does it feel harder or easier?
  - Are you using the same muscles as before?

- Some people get a cramp the first few times, do you?
- Compare how it feels by lifting each leg with a flexed foot and then with a pointed foot.

- Point your foot and bring the knees toward one another as you lift each leg up. What feels different now? Bring each leg up with the knees out and then with the knees in. Notice what feels different.

- Continue lifting each leg by pointing the foot and bringing the knees together, focused on releasing every muscle you feel in your torso.
  - Are you contracting the buttock muscles?
  - Is your foot of the opposite leg pushing into the floor?
  - Is your lower leg and knee contracted or stiff?
  - Are you bracing with the opposite leg, buttocks, neck, shoulders or chest?
  - Release all that you can feel as you lift each leg. Is either leg getting lighter to lift?

- Finally, as you lift a leg feel what is contracting in the leg itself.
  - Is it contracted in the back of the leg?
  - Is the calf tight?
  - Is your ankle stiff?

- Feel how to release muscles in both legs and the torso as you lift. Try bringing both legs up as you feel how to release the pelvis, letting the chest slump and your head hang down with the chin tucked in (Fig. 28).

- Stand up and notice how it feels to lift each leg in walking. Do they feel different? Notice as you lift a leg in walking whether you hold your breath, stiffen the opposite leg, contract and lift your chest, neck, and belly. If you do, stand in one place and lift one leg and then the other feeling how to release throughout your body just as you did on the chair. Once you release every muscle you feel, try walking.

Are you surprised at how involved your upper body is when lifting a leg? When the pelvis becomes immobile your brain tries to use your upper body muscles to lift the legs. Once your pelvis moves freely you must retrain the brain to *not* contract the upper torso to lift a leg. Pushing through the feet in sitting also affects the way you come to standing. Try the following sequence of sit to stand movements:

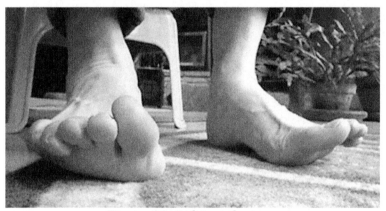

Figure 29. Lift just the toes

- As a baseline stand up and notice how you do it.

    - Do you lock your knees?

    - Does your chin come up and your neck stiffen?

    - Do you stop breathing or contract the belly muscles?

    - Do you tighten your buttocks or round your shoulders?

- Do you use your hands to push off?
- Do you initiate standing with your head and chest?
- Is your weight in your heels? Sit and stand a few times to feel what you do.

– Sit with the legs at a 90-degree angle and hip distance apart. Press your feet into the floor. Notice what contracts. With the feet pressed into the floor try to stand up.

Figure 30. Rock forward and rest on the legs

- Does your weight go into the heels?
- Do you need to use your hands to push off?
- Does your pelvis roll back and feel stiff?
- Is it harder to stand?
- Do you pull forward with your chin and neck?

- Sit and lift just your toes (Fig. 29). Does your weight go into the heels? Feel how to lift the toes as if your knees are being pulled forward. Do you feel less pressure in your heels? Lift the toes again as if the knees are being pulled forward. Does your pelvis begin to roll forward with the pull? Does your chin come up and the chest lift? Does your lower back feel tight and arched? Lift the toes by 'pulling' the knees forward until it feels light and easy and your feet rest lightly on the floor.

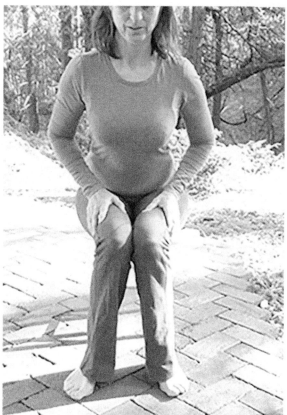

Figure 31. Fall forward with straight spine and knees together

- Tuck your chin in and down so that the back of the neck lines up with your spine and feels elongated. Release the chest without rounding your back. Slowly allow the torso to fall forward toward your thighs. Do you push through the feet? If so, lift the toes as you fall forward. If you are able, allow the torso to be flat on your thighs (Fig. 30), if not put the elbows on your thighs to support the torso (Fig. 26). Your back should feel flat and long all the way up through the back of the neck. You should also feel pressure on the front of the sit bones as the pelvis rolls forward like a ball.

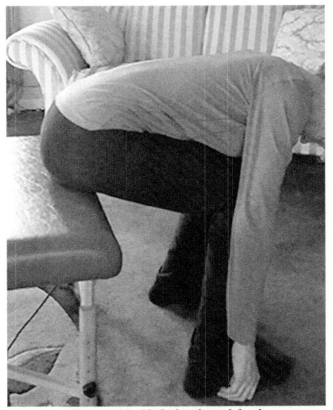

Figure 32. Slide back and forth

- Let your knees "fall" toward one another until they touch and slowly repeat the movement (Fig. 31). Only go down so far as you can allow the knees to stay together. Are you pushing through the feet again? Is your pelvis stuck, unable to roll forward? Is more weight on the inside of your heels with the knees touching?

- Tip the inside of the feet onto the floor and let the knees "fall" together. Is it easier to let the knees be together? Try the movement in this configuration. Did the pelvis free up a bit? Is the weight still heavily on the inside heels? As you bend forward further does the weight go toward the inside ball of the feet. Does this free the pelvis up even more?

Figure 33. Slide forward till pelvis launches

- Tuck your chin in and slightly down and take the head back until it aligns vertically with the rest of the spine. Lift the outer toes to tip the feet inside, then lift the rest of the toes and allow the knees to fall together. Let your pelvis roll forward until the legs stop the "fall". Let the arms dangle down in front of you. Notice how difficult it is to breathe from the belly in this position. Can you breathe from the diaphragm? Repeat many times until you release all that you feel. As you practice, pause when bent over so that you cultivate diaphragm breathing.

- Sit again, with knees together bend over until the torso rests on the legs. Let your torso drop even lower so your pelvis slides forward on the chair. Catch the forward movement with your arms on the floor. Feel how to slide the pelvis back and forth on the chair (Fig. 32). Each time you slide forward and back feel how to further release the knee and ankle joints. If your pelvis slides off the chair you have gone too far. Notice if lifting your toes makes it easier to slide the pelvis forward. Can you feel how to press through the inside ball of the feet to slide the pelvis back?

Figure 34. Straighten legs, spine lengthened

- This time bend all the way over your legs and stay there. If the chair is higher than the length of your lower legs and feet, the torso will automatically slide forward on the chair. If the chair is lower, let the torso drop lower until you slide forward on the chair. As you slide forward let your ankles release and you should feel, much like a plane taking off, when the pelvis naturally raises in the air, automatically beginning to straighten the legs (Fig. 33). Notice if you brace through your legs as the pelvis slides off the chair. Does everything stiffen as soon as the pelvis is not touching the chair?

Figure 35. Straight spine to stand up

- This time, bend over and slide the pelvis forward until it gently launches in the air, naturally straightening your legs. Leave your torso hanging down and feel how to lower the pelvis to the chair and gently re-launch it (Fig. 34). Repeat many times feeling what muscles you can release. Keep your arms in front of you so that even if you do fall forward, your arms will catch the fall. Repeat the movement until it feels easy.

Figure 36. Bending over with a frozen pelvis

- Once you can comfortably stand with the pelvis in the air, legs straight, and torso hanging, tuck your chin and feel how to bring the torso upright (Fig. 35). The movement should occur at the pelvis. Do the feet press into the floor again? Let the torso come back down the same way you brought it up. Release your toes up slightly and let the knees release forward so that your weight is in the balls of the feet. Staying forward, come up to stand with your chin tucked and back straight. Do the feet press into the floor with less force? Repeat the movement until it is effortless.

It is far easier to stand up when your pelvis behaves like a ball allowing the upper and lower body to be separate. In this sequence you also learned how to come from sitting to stand by releasing your body in gravity through yielding rather than resistance, contraction, and bracing. The movements must be exaggerated in the beginning so that you can feel what needs to change. To sit down is to simply reverse the order of movements. Eventually you will feel *lightness and balance* as you effortlessly come to stand from sitting, or from standing to sit.

One of the most common irritants clients complain about is tight hamstrings and lack of flexibility. The ability to be flexible has everything to do with your skeleton, specifically your joints. If joints feel stiff because muscles are in conflict as you move, then you are likely to be less flexible. Notice the difference between two techniques of bending over in Figure 34 and Figure 36. Compare how close the hip points are to the top of the thighs. Notice how the upper spine is flexed in Figure 36 and straight in Figure 34. Are the legs straight or bent, and does the weight appear to be in the heels or the balls of the feet? In figure 36 the arms are further from the floor which shows what happens when contracted muscles constrain the pelvis from rolling forward and letting the torso drop down. The brain tries to create the torso bend by collapsing the chest and flexing the upper spine. Try the following practice to improve bending:

- Stand with your feet hip distance apart. Allow the pelvis to roll forward by releasing the belly and low back. Tuck your chin and let the torso move with the pelvis. Release the belly muscles as you bend.

- This time tuck the chin, keep the spine long, release the pelvis forward and roll your knees toward one another so that they cannot lock or hyperextend back. Let your weight go into the balls of your feet by lifting and/or releasing your toes. Keep releasing all muscles of your torso. Are your hands closer to the floor? Keep releasing every time you repeat until your hands touch the floor.

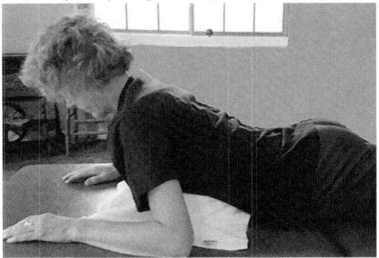

Figure 37. Lifting the head with dropped shoulders

Your head is organized to move on its axis. The neck part of the spine can move somewhat independently from the rest of the spine because of a connecting vertebra called C7. Above C7 the neck can move in all directions as the head counterbalances on its axis. If the neck and head do not differentiate from one another the torso becomes top heavy and compels us to use our chest and shoulder muscles for control. The top-heavy torso tends to pitch forward which causes the belly muscles to contract to prevent the body from falling forward. The legs become contracted and stiff which immobilizes the pelvis from below and above. By relearning how to balance the head on the spine and turn from its axis you reverse the balance issues that make you pitch forward. Try the following practice to feel how to balance the head on the spine and turn from its axis:

- Lie face down on the floor with a firm cushion under your chest such that the chin hangs over it. Bend the elbows out with your hands toward the chin. Tuck the chin and begin to lift your head, keeping the chin tucked. Notice if the shoulders hunch and your chest comes away from the cushion. Do you stop breathing?

- Try the movement again, only this time drop the shoulders down (Fig. 37). What happens to your chest? Does your breathing change or are you still holding your breath?

- Once again, lift, drop the shoulders, and release the chest down. Does the head feel lighter to lift? Do you feel more like a snake? Is the breathing changing?

- Each time you lift by releasing the shoulders and belly muscles the head should become more upright as the belly and pubic bone melt into the floor. The spine becomes more arched and you feel more like a python. Repeat the movement and notice if you can begin to slightly turn your head from side to side without involving your neck, shoulders or torso (Fig. 38).

Figure 38. Turning the head on its axis

I saw a child who lost the ability to control or lift her head at all due to a neurological condition. She had cleverly figured out how to keep her head upright in sitting by hunching her shoulders to cradle it. However, what she needed to relearn is how to lift and support the head from her spine. Muscles that were trained to compress in the front needed to be retrained to release and lengthen thus freeing up the spine to arch and curve to lift the head. Once upright, she re-learned how to move her head on its axis without losing control or balance.

Differentiation allows us to move with an elegance and grace that is pleasant to feel and looks more attractive. Fully differentiated, we move fluidly in well-oiled flexible parts that seamlessly coordinate the entire body in any activity. Movement is effortless allowing the conscious self to process new and different data rather than being distracted by discomfort, pain, and limitations. In the next chapter differentiation is integrated into the ability to rotate as we move.

7 - Element 4: Rotation

An object is chiral if it cannot be brought into congruence with its mirror image by translation and rotation. - Vladimir Prelog

Rotation and counter-rotation is central to the way the body maintains its center of gravity as you move. According to physics an object is balanced in its center vertically half way between the top and bottom, and horizontally at the center, where it crosses through the vertical balance point. In the human skeleton, vertical and horizontal halfway points are in the middle of the pelvis (Fig. 39).

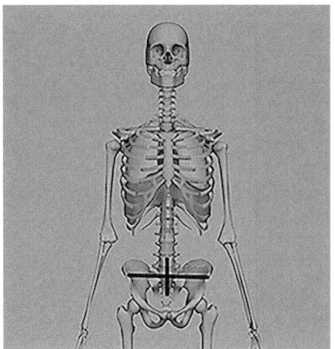

Figure 39. Center of gravity

Given where the center of gravity is, if the part of the body above the center (the torso) rotates to the right, the part of the body below the center of gravity (the pelvis) must equally counter-rotate.

If no rotation occurs, then the body will move from side to side like a box with feet on each corner.

A box is supported by the walls on the outer edge of the structure. To get an idea of what I mean walk towards a mirror. Notice if your body goes slightly side to side. Do you feel weight on the outside of the feet, legs and hips?

Compare the two walks in Figure 40. The walk on the left is from side to side. Notice how the weight transfers to the outside of the legs and feet. The bend in the waist shows no spine support above the hip on the outside, so the outside muscles must support the torso. Most people lose their waist by the time they are middle-aged, and nowadays even young people are losing their waist because they do not move from their center. When we move from side to side the brain sensory map becomes two-dimensional as we have a sense of front and back only.

Figure 40. Walk comparison

The person on the right (Fig. 40) walks such that her weight goes through the center of her bones and joints, including the spine. You can see the spine twisting between the upper and lower body which means the muscles are working through her center. When moving from our center the brain sensory map is three-dimensional. Neil deGrasse Tyson explains three-dimensionality by visualizing a black hole. He said, "You need to imagine the hole is in the middle of a room so that no matter where you are in the room you are looking toward the center of it." Every living thing on earth, including the earth itself, is based in roundness. The most efficient way to move a round object is by rotation from its center. To experience circular motion, grab any size ball and notice how much effort it takes to roll it. A ball easily rolls in any direction from its center.

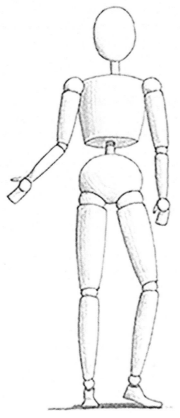

Figure 41. Cylinder body

Isaac Newton's second law of motion married with non-uniform circular motion formulas[29] is mathematically how humans are designed for optimal efficiency in movement. The body can best be pictured as a series of cylinders (Fig. 41). Each cylinder moves independently of other cylinders through joints. To balance from the center of gravity, each cylinder component must rotate in one direction while the adjacent component equally counter-rotates in the opposite direction. For example, in a walk step the upper leg rotates slightly toward the center of the body while the lower leg counter-rotates equally in the opposite direction. This insures the components stay in the center of the knee joint that connects them. By lying on the floor to practice turning you can safely get a sense of how your body organizes itself to walk:

- Lie on your back, arms and legs long, and begin to initiate a roll to your left side. Is it difficult to initiate? Do you need to leverage by pushing parts of the back of yourself into the floor to roll yourself left?

If you push parts of yourself into the floor to turn your brain perceives the body as a two-dimensional box. Figure 42 shows how difficult it is to end up on a side if the brain perceives the body as a box. The march of the cards in the Disney film Alice in Wonderland[30] illustrates movement in two dimensions. All movement is from side to side. Of course, in the Disney animation when a card falls on its front or back it magically floats upright, but in the real world it takes a huge effort to roll a box onto its side. Try the following to see how you can help yourself to become more round to roll onto a side:

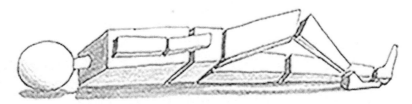

Figure 42. Box body

- Bend your right knee so that your foot is standing on the floor. Drop the knee to the left and try to roll. Is bending the knee helping? Press the right knee over the left knee toward the floor. You can use one of your hands to press the knee left. Are you beginning to roll left? Do you still brace parts of the back of yourself into the floor? It's a lot of work to push forward from your back because you are pushing your body weight against gravity!

- Bring the right knee diagonally across the chest toward the left shoulder. You can assist with a hand. Is it any easier to roll? Do you tend to slide left rather than roll? Do you still brace through the back of yourself to roll?

- Bend the right knee and drop it over the left leg. Use the floor to press through the inside of the right foot to roll left. Notice where you block the movement. Does your chest and chin come up? Release through your neck, shoulders and upper back.

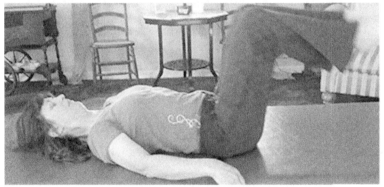

Figure 43. Knees up rounding the back

- Bend the legs up slightly and roll the knees together. Can you feel the muscles above your pubic bone working? Keep rolling the knees into one another as you release your buttocks, back, chest, shoulder and neck. Repeat until you feel as if only the muscles in the upper inner thigh and muscles just above the pubic bone are working.

- Draw the knees toward your chest keeping them rolled into one another. As you draw the knees up feel how to release the chest down and lengthen your back. Repeat the movement by pointing the feet with the toes turned out, squeezing the knees, and sucking in the belly muscles (Fig. 43).

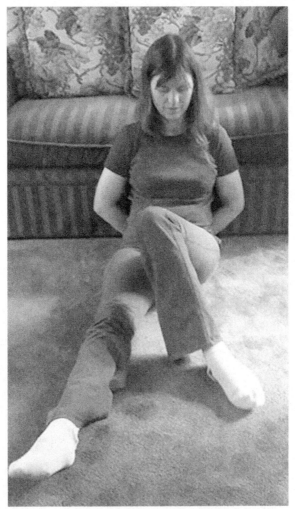

Figure 44. Bring leg up, knee in, foot turned out

- As you bring the knees up this time allow them to fall to the left. You should not brace with any part of the back of the body. Feel how the knees falling left pull the rest of the torso to the left side without needing to use muscles or force. As you refine the practice of rolling to your side by bringing the knees up you may begin to feel more like a ball. The easier the movement becomes the rounder you feel as you initiate from your center of gravity instead of the outside edges of your body.

If you don't rotate and counter-rotate as you move wear and tear on joints increases as pressure is unevenly distributed. Uneven pressure often leads to low back pain, sciatica, stiff neck, knee and hip problems, and more. Once the brain perceives the body much like a box movement is from the edges of the skeleton. Most adults walk or run with no rotation or counter-rotation through the legs. The legs bend, but they function like a hinge.

Try the following to see how you bend your legs:

- Sit with your back leaning against a wall. Slide your buttocks forward so that your torso is slumped. Bring one leg and then the other up toward your chest. Does your knee come straight up toward the chest or does it angle out? Do you lift the toes or use the foot to bring the knee up? How far can the knee come toward your chest without forcing it? How hard does the movement feel? Does the chest lift or the pelvis roll forward?

- This time point your left foot, turning the toes out slightly. Slowly roll your left knee toward the opposite knee while keeping the toes pointed slightly out (Fig. 44). Does your knee naturally start to come off the floor? As you repeat, release your buttocks, chest, shoulders, and neck. Does your back press into the wall? Keep slumping the torso down and forward as you rotate the leg. Make sure no part of your body pushes into the floor as you do the movements.

- Imagine that a string attached to the leg just above the knee is pulling it up at an angle toward your center. As the knee pulls up the toes stay pointed out. Release all muscles in your torso, allowing the chin to drop down and the spine to flex. Notice as you repeat the movement that the upper leg rotates toward your center while the lower leg rotates away from your center. Repeat many times, feeling as you release muscles that the leg becomes lighter and the knee more flexible.

Figure 45. Both knees up together

- Bring your left leg to standing. Let your left knee drop over the right leg and begin to draw the knee up toward your right shoulder. Keep the toes turned out as you draw the knee up. As you slide the knee up and over to the right your head and torso sink down toward the pelvis. The spine goes into flexion as if you are rolling into a ball. Make sure no body part pushes into the floor as you do the movement.

- Roll your knees together with feet pointed and toes out. Lift both legs toward your chest, making sure your chin and chest stay collapsed down (Fig. 45). Feel the bottom of the pelvis roll up toward your chest. Let the front of your body collapse as if you are rolling into a ball. Don't reach down with your chin, but allow the spine to curve like a "C".

- Bring the knees to your chest by rolling the pelvis up and hold. Roll the knees firmly together and let them drop to the left. The upper back stays on the wall while the pelvis must diagonally cross to the left. Feel how to release your ribs, chest, shoulders, and neck. Repeat on the opposite side.

How do these movements translate to walking? To practice use a wall to lean against so that you need not be challenged with trying to maintain your balance:

Figure 46. Leg organization to walk

- Stand facing a wall about a foot or two away from it. Place your hands on the wall below shoulder height with your elbows slightly bent in toward your center. Turn the toes of both feet out very slightly. Put all your weight into the inside ball of the left foot and slightly rotate the right knee toward your center. Allow the right knee to release forward rotating slightly toward your center. Let the right heel come up but do not lift the toes off the floor. The angle of the foot stays slightly turned out so that the part of the foot touching the ground is on the inside ball of the foot. Release the toes. Let the knee fall toward your center by releasing the heel up. Once the heel is as high as it will go imagine someone just pressed the back of your knee forward (you can also get a partner to push the back of the knee forward) and feel how to release the leg forward. As it goes forward lift the knee diagonally toward the opposite shoulder. Keep your chin tucked and chest released (Fig. 46).

- As you repeat the movement make sure your upper torso stays squarely facing the wall. You should feel that the right side of the pelvis is angling diagonally toward your center. Repeat the movement until the leg feels light and springy and your pelvis moves without affecting the upper spine.

- Once the movements are light and easy, try stepping back from the wall and feel how to release one knee forward while balancing through the center of the opposite foot. Your pelvis should feel as if it is dropping diagonally forward toward your center. Stand in front of a mirror while doing this practice, to ensure the pelvis is not going side to side (Fig. 47).

How we move can either go with gravitational force or against it. One way to move against gravity is to push an object. For instance, the earlier practice of rolling to your side demonstrated push versus pull. To go with gravity, movement is initiated as if your body is being pulled. Try the following movements to see if you can feel what it is like to go with gravity:

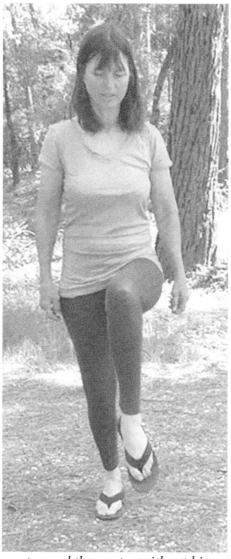

Figure 47. Knee up toward the center without hips going side to side

- Face a wall 1-2 inches away with feet hip distance apart, toes slightly turned to the outside, and heels toward your center. Let your forehead and pubic bone drop forward to rest on the wall (Fig. 48 left). If your knees bend, straighten them. If you cannot straighten your legs and keep your heels down, move closer to the wall. If the belly prevents your pubic bone from touching, try sucking it in. Do you contract your neck muscles? Do your hamstrings, buttocks and calves feel tight? Repeat the movement many times, feeling what to release so that the pubic bone rests easily on the wall.

Figure 48. Line up the structure on a wall

- Let the pubic bone and forehead come forward to rest on the wall. Slowly push just your torso back from the wall until you are balanced in a freestanding position (Fig. 48 right). Do not bend or lock your knees. Let your hands release from the wall and feel how to adjust your balance in space; you may feel quite different from your usual standing posture. Feel how to let just the pelvis (lead with the pubic bone) come gently forward in space until you touch the wall. Release your ankles and lift your toes as you fall forward. Make sure your upper chest and head remain vertical as you fall forward. Repeat the movement many times until you feel more comfortable letting the wall catch you.

Figure 49. Let the wall catch your fall

- This time stand 6 inches to a foot from the wall and let the pelvis fall forward into the wall (Fig. 49). Allow your left knee to come forward by releasing the left heel up. Make sure the toes are out and the heel is in toward your center. Release the heel up to allow the left knee to drop forward and toward your center, pulling the left side of the pelvis forward at the same time. Make sure your weight does not drift to the outside of the right foot. Keep your upper body vertical as your pelvis falls into the wall. Catch the fall with your hands while keeping the upper chest and head vertical. Do not lift your chin or chest as you repeat the movement. With repetition, you should begin to feel how to coordinate the knee lift with the pelvis so that the fall forward is controlled and reversible within the movement.

Figure 50. Take a step back while leaving the pelvis forward

- Fall into the wall as before and push just your torso off the wall, take a step or two back and lift the left knee toward your center with the toes pointed out and the heel pointed in toward your center. As you step back leave the pelvis forward so that you feel just your leg stepping behind you yet your pelvis and torso remain forward (Fig. 50). Notice as the pelvis comes forward how the torso must lengthen and become more vertical to keep you in balance. Do not hollow the chest or take the torso forward in a mistaken attempt to prevent you from going backward. Make sure the knee of the supporting leg is slightly turned toward your center, the knee is unlocked, and your weight is on the inside ball of the foot.

Figure 51. Walk with the pelvis leading

- Push off the wall, take a step or two back, turn and take a step forward, letting your pelvis lead the way (Fig. 51). Notice as the pelvis comes forward how the torso lengthens and becomes more vertical to stay balanced. Do not hollow the chest as you lift one leg and then the other leg. The supporting leg should be unlocked and weight on the inside ball of the foot. Face a mirror when you transfer weight from one leg to the other to make sure you do not swing the pelvis from side to side.

By leading with the pelvis in walking your torso adjusts itself to be more upright and over the pelvis to maintain balance in your center. Does walking require less effort as you feel gravity working in your favor? Notice when you step forward or step back that you balance quite similarly as the pelvis remains centered over the front of the feet.

Turning to look around yourself requires the head and neck to move independently of the rest of the torso below. The head is designed to turn independently of the neck on its axis. The neck has some independence from the rest of the spine at C7. Most adults I see have no neck or head independence and struggle greatly with turning to look around themselves. It makes driving hazardous as people rely on mirrors and less on turning to look over their shoulder. In the interest of less accidents and close calls on the road, we must all be able to safely look around ourselves. The following practice helps you feel how to turn using your axis and neck independently of the torso to look around yourself:

- Sit with your feet hip distance apart, knees dropped toward one another and, if possible, resting against one another. Allow the pelvis to roll forward just enough to lengthen your spine. Lengthen the back of the neck by tucking your chin slightly down toward your collar bone. Slowly start to turn your head to look to the left, keeping the chin down. As you turn do not let the legs fall left or right (Fig. 52). Do not lift the chest. Each time turn just your head and feel what muscles to release in your neck and shoulders.

Figure 52. Turn to look without rotating the pelvis

- This time turn just your head and pause a moment. Release muscles you feel and continue turning the head. Can you feel the neck starting to turn as your head reaches the limits of its axis? As you continue slowly turning with the chin tucked feel what to release in the belly, legs, chest, shoulders and arms. Imagine a pole down the middle of your torso from the head that you must twist around. Repeat many times releasing what you feel.

Figure 53. Turn using the upper body only

- As you turn the head and neck notice if your chin runs into the shoulder. Allow the shoulder to turn to move with the chin. You may feel where the neck starts to engage the rest of the spine as you continue turning left to look around yourself. Make sure you do not lean into your right sit bone. You should feel at some point the involvement of your spine as you rotate. Keep the legs centered and do not allow them to drop left or right. Keep imagining a pole is down the center of your torso and you are constrained to turn around it. Turn your head only as far as your upper torso allows (Fig. 53).

Release whatever muscles you feel as you repeat the rotation many times.

- Does your left side feel different from your right side? It may feel bigger and rounder. This is your brain updating its sensory information, which is just beginning to perceive your body as round. Lie on the floor and sense the distance between your front ribs and the ribs in back of you, on the left side and then the right side. Is there any difference in the distance between the two sides?

- Repeat the turning sequence on the opposite side.

One more element is critical in how we walk, run, jump, and go forward. In the next chapter propulsion is integrated into your movements.

8 - Element 5: Propulsion

Way before we were scratching pictures on caves or beating rhythms on hollow trees we were perfecting the art of combining our breath and mind and muscles into fluid self-propulsion over wild terrain. – Christopher McDougall

Newton's third law of motion, the organization of the skeleton, and tendons are what propels the body in any direction with explosive power. Propulsion makes it so that you can jump higher, run faster, and have more stamina and power. Watch any animal run and you will see their head does not bob up and down, they thrust forward with each stride, but their head must be stable so that they can keep their eye on where they are going, or on what they are chasing. Walk a few steps and notice if your head goes up and down. Run a bit and it will be more noticeable. When we move with true propulsion the head does not bob up and down. How must the skeleton organize to move without bobbing up and down? Rotation and propulsion are the keys to smooth, gliding, light, powerful movement. In walking or running where does propulsion originate and how does it work? First, you must feel how you habitually walk:

- Walk a few steps. What creates forward momentum? Is your head in front of your pelvis? Do you lift your feet by bringing your toes up? Do you land on the outside of your foot as you step forward? Does your belly contract out? Do you stiffen your neck, chest and shoulders? Does any part of your skeleton rotate?

- Slow the walk down and notice how you place the leading foot on the ground. Do you land on the heel and travel through to the front of the foot? Or does the foot land quite flat? What does the foot behind do? Do you lift the back foot as a single unit? Does any part of it leverage into the ground? Does your back foot play any role in sending your body forward? Does your head bob up and down as you walk, or do you walk with your head forward and down? Do you lock the knee of the supporting leg at any point in the walk step?

Figure 54. Leverage the foot against the ground to move the pelvis away

The feet play a critical role in propulsion for walking and running, but few adults initiate movement from the foot. The following practice should clarify how to leverage through the foot to move other parts of the body. In addition, it will give a sense of how rotation and propulsion are integrally related to the movements:

- Sit on a flat surface with your feet hip-distance apart. The height of the chair or bench is best if the legs are at a right angle to the floor. Place your right foot to the side and under you with the top of the foot on the floor. If your foot is on a hard surface put a yoga mat or piece of rug under it. Drop your head and chest down but keep your torso upright. Do not lean on the surface with your hands. Gently press where the front of the foot meets the floor (Fig. 54). Feel how to release the heel, back of the calf, back of the upper leg and buttocks. Release the opposite leg, belly, ribs, chest and neck. Let the chest and head stay dropped forward but upright. Release the inside of the upper thigh, belly, buttock and back muscles as you repeat the movement many times. As you release the muscles you should feel how the movement affects just the pelvis while the torso yields and balances through the spine.

If your pelvis and torso are moving as a unit you are contracting the belly muscles. Contracted belly muscles block energy from passing through when you walk, run or jump making it more difficult to get any propulsion from the movement. Try the following to release the belly:

- Lean your hands on the chair and round your back. Release the belly muscles as you push through the top of the foot. Is the pelvis beginning to move on its own? If not, push through the arms as you push through the top of the foot. Can you feel where you block now? Try to release all muscles in your torso as you repeat the movement.

- This time don't push through the arms, push only through the top of the foot. Does your pelvis move on its own now? If so, straighten your torso and push through the foot. Can you move just the pelvis without the rest of the torso moving forward or sideways?

Figure 55. Leverage the foot to move the pelvis forward

By pushing through the top of the foot you are constrained from pushing through the heel so that you can feel the front of the leg work rather than the back of it. Once you feel that the sitting practice is affecting just your pelvis, try the same practice in standing:

- Stand 1-2 inches in front of a wall with your hands supporting you. Let your pubic bone and forehead rest against the wall. Place the left foot behind you with the front of the foot on the ground. The foot is pointed with toes out. Drop the heel toward your center. You should feel the joint just behind the top of the big toe touching. Make sure the supporting leg is straight but not locked with a slight rotation of the knee toward your center, toes slightly turned out, and weight on the inside ball of your foot. Press gently through the joint just behind the big toe to feel how it passes through your skeleton. Notice as you push that the right side of the pelvis presses into the wall as the left side rotates back. The left leg lengthens as the heel comes further up. Release the heel, calf, back thigh, buttocks, low back, belly, ribs, shoulders, and neck as you repeat the movement many times. Keep your chin down and notice as you press through the front of the foot whether you feel the torso lengthening. Is your spine lengthening through the back of the neck? Release your lower jaw forward and down (Fig. 55).

- Breathe from the diaphragm as you repeat the movements. Notice if the upper chest goes toward the wall as you breathe. Suck the belly muscles in so they cannot be used in breathing.

Once you can feel how to move your pelvis forward by pressure through the top of the foot, you are ready to try using the bottom of the foot in the gap just behind the toes.

- Stand facing a wall, 1-2 inches away with your hands against the wall at shoulder height, and your pubic bone and forehead resting against the wall. Place one foot behind you as a sprinter would place it on a running block, with the toes flat on the ground and the heel vertically pointed toward the ceiling. Let your inside heel drop toward your opposite leg. The back of the leg and buttock are relaxed and released. Make sure the supporting leg is straight but not locked, with a slight rotation of the knee toward your center, toes out, and weight on the inside ball of the foot.

Figure 56. The ball of the foot propels the pelvis forward

- Imagine your toes are attached to the floor with Velcro as you begin to leverage against the floor to propel your pelvis forward (Fig. 46). Press the toes into the floor on purpose and then release them. Press through the gap after the toes to move the pelvis forward. Allow the leg to lengthen and the heel to lift higher and come forward as you push, releasing the back of the leg, buttock and back. As the heel comes higher you should feel more pressure on the front of the foot behind the toes.

- As you lengthen the front of your body, do you feel taller?

- Does the movement push the right side of the pelvis into the wall?

- As the leg straightens, is the left side of the pelvis pulled back? The pubic bone should stay against the wall on the side that is rotating forward.

- Does your chin drop down slightly as the back of the neck lengthens?

- Keep your upper torso pressed into the wall. You should feel the chest press into the wall even though the collarbones are lifting up and back toward your neck. As you push, make sure you do not bend or lock the supporting leg or push it into the floor. Breathe from your diaphragm feeling how each breath lengthens the front of your body even further.

- Repeat the movement many times, feeling how to release muscles throughout your body so the pelvis rotates with greater ease and range of motion. The goal is to feel that the pelvis can move completely independent of your upper body.

Once you feel comfortable with the movement and there is almost no weight through your hands, try it in open space. Adjust and release what you feel so that you remain balanced as you push through the gap behind the toes, allowing your pelvis to move forward (Fig. 56).

Propulsion is how we move the body. It works best when no muscles block energy from traveling through the skeleton. The next chapter integrates all five elements into a common activity to feel how they work together so that you move as your body evolved to function.

9 - Putting it all Together

Movement is the action that connects your body to your mind in the way that you were born to function.

In this chapter I show how all five elements are incorporated into a walking practice. Walking is a highly complex activity. It takes children about a year to learn to walk, as they progress through a series of pre-walking functions that include lifting the head, rolling over, sitting up, putting weight through the feet, crawling, and, finally, standing up. Every pre-walk movement sequence they learn is completely reversible so that when they finally do walk, automatic adjustments are made in every direction to maintain balance. Most of us don't think of walking as being just as ready to move backward or sideways as to go forward, but centered balance in movement is the ability to go in any direction with ease. Try walking backward or sideways and notice how comfortable, balanced, and easy it feels. To walk backward requires more attention and awareness because you don't do it much. It is like brushing your teeth with your left hand if you are right-handed or vice versa. The following practices show you how to walk backwards in balance as if you are just as ready to walk forward, which is why walking backwards, in a non-habitual way, helps you improve your forward walk:

- Stand up with your feet hip-distance apart, toes slightly turned out and heels slightly in. Allow your pelvis to come forward in space until you are solidly on the balls of your feet. Slightly turn your knees toward your center. Breathe through your diaphragm, naturally allowing the chest and rib cage to lift. Tuck the chin in and slightly down so that the neck is vertically aligned with the rest of the spine. Lift your left foot and place the ball of the foot 1 - 2 feet behind you leaving the heel up (Fig. 57). The left side of the pelvis rotates back while you keep the right side forward. Your upper body should stay facing the front. Allow the left heel to go toward the ground. Keep the right side of the pelvis facing forward as the heel goes up and down. Can you feel a twist in your spine, as the heel goes down? The upper chest stays facing the front so the twist is felt lower in the spine. Does the spine movement slightly arch the ribs forward and up?

Figure 57. Pelvis goes back diagonally

- Repeat the movement of bringing the heel up and down many times with the left leg. Pause. Now repeat the entire sequence with your right leg.

- In standing lift the heel of the right foot and let the knee drop in toward your center. Lift the leg as if a hook is attached to the top of the knee allowing it to travel forward toward the center line of your body. Swing the right leg back and place the ball of the foot on the ground 1-2 feet behind you and hip-distance apart. The right side of the pelvis should rotate back and allow the leg to *swing*. Do not lift your rib cage, stiffen your left leg, or hold your breath as the right leg swings back. Make sure your heel is toward your center as you place the ball of the foot on the floor (Fig. 58). Keep the torso forward. Let the heel of the right foot touch the ground while keeping the pelvis and torso forward. Keep the chin down throughout the movement. Repeat several times and then switch and repeat the sequence with the left leg.

Figure 58. Leg swings front and back

- This time swing the left leg back leaving the heel up and weight on the ball of the foot. Initiate with the pelvis to rotate and swing the leg to the front. Lightly place the heel on the floor in front and then rotate the pelvis to swing the leg back again, landing on the ball of the foot. Repeat the swing back and forth many times making sure to keep the torso centered and facing forward. Do not swing from side to side. Imagine there is a pole down the middle of the pelvis and you must rotate only around the pole. Repeat the movement back and forth, slowly at first, and then faster and easier noticing what more you can release. Do you feel the organization of your balance is the same whether your leg is behind you or in front of you? Is it reversible? Reversibility of movement means you can swing the leg forward or swing the leg back without changing your balance.

Any movement is reversible if it is balanced through your center. The key to a good partnership in couples dancing is that everyone is balanced within their own body and ready to follow or lead, in any direction instantaneously, as the dance dictates. When you step backward, the upper torso goes forward, and when you step forward, the upper torso goes forward. Walking backward is the same as walking forward to the torso because it must always be forward to stay balanced in the center.

- Practice a leg swing back and forth a couple of times. On a backswing let the front part of the back foot bear weight (Fig. 58). As you let the heel lower keep the weight in the ball of the foot. Do not let weight go into the back foot heel. Does your torso lengthen and become more vertical?

- Does your spine lengthen through the back of the neck as the chin is pulled slightly down and in? As the leg swings forward the heel meets the ground in front first. Let your pelvis and torso come forward as weight travels toward the ball of the front foot. Does your spine lengthen? Does your front naturally lift while the chin stays down and aligned with the spine?

Figure 59. Feel the pull forward

- This time push slightly through the ball of the foot to propel the front leg backward. Notice how the pelvis rotates slightly diagonally as the back leg meets the floor. Allow the toes to release into flexion so the rest of the foot meets the floor. The heel of the back leg should never feel weight in it, even if it is on the floor. Keep your upper body forward so you do not fall onto the back heel.

- Repeat the movement of stepping back several times. Try stepping back, then forward, starting to step back but deciding to step forward instead and vice versa. Notice if it feels easy and balanced to "change your mind". Make sure your pelvis leads in every movement.

- This time with the left leg back, feel how to push through the ball of the foot to propel your pelvis forward. The heel should release up as you leverage through the inside ball of the left foot. Notice how the push transfers weight from the back leg through the body to the leading leg. The pelvis is propelled forward bringing the upper torso with it such that the spine lengthens and becomes more vertical. The back of the neck lengthens; the chin drops down and in toward your collar bone. Your head should glide forward with no up or down motion. Once you feel how to push through the ball of your foot to propel your pelvis forward and transfer weight to the opposite leg, let the heel of the back leg keep coming up and forward until the leg *pops* up and forward into your next step. Let your chest and belly release while keeping the chin slightly down and tucked. Make sure you do not lock the knee, ankle or hip of the standing leg. Repeat the movement many times of eventually popping the back leg up and swinging it forward without placing it on the ground.

- Repeat the movement only this time let your heel touch the ground while allowing the momentum to pull your upper body forward. Feel how your upper body becomes more vertical and the spine lengthens (Fig. 59). If the heel is not turned slightly toward your center and the toes turned slightly out, you will not feel the pull forward from the center of your pelvis, but rather a turning from the outside as if your body will turn from its outer edge. As you glide forward notice at what point in the movement it seems right to push through the ball of the back foot to continue propelling your body forward in space. There is a brief moment when weight is transferred into the front foot making it the supporting leg so that the opposite leg is free to swing through in your next step forward. The supporting leg transitions to be the back foot propulsion that becomes your next step. Repeat the sequence many times feeling how to glide across the floor with no abrupt, jerky, or up and down motion.

Learning to move from your center will reshape muscles around the skeleton to create a more attractive, symmetrical, and elegant appearance to your body. The waist gets smaller, thighs and hips slim down, the shoulders get wider and more square, and the neck looks longer as it slims down. Many distortions from muscle misuse such as loss of height, scoliosis, bow-legs, thick ankles, big bellies, and more disappear as you learn to move the way your structure evolved to function. Walking has the potential to become sinuous, balanced, smooth, effortless, and attractive. Studies on body language show you have about thirty-seconds to make an impression when you walk into a room. Olivia Fox Cabane, in her book The Charisma Myth tells the story of a reporter that spent a day with Marilyn Monroe traveling around New York City. The reporter was surprised that no one recognized her as they traveled about. Toward the end of their time together he asked her about it. She gave him a wide-eyed look and offered to "turn it on" for him. They had just come upstairs onto the sidewalk from the subway when she tossed her hair and changed her stance, and to his utter amazement people immediately began flocking round. It quickly escalated into a mob scene and they barely made their escape into a cab. Marilyn had learned, through small movement changes that transformed her presence, how to turn on the *charisma zone*.

Consider this poem, "Phenomenal Woman" by Maya Angelou in her book, <u>I Know Why the Caged Bird Sings</u>:

I walk into a room
Just as cool as you please,
And to a man,
The fellows stand or

Fall down on their knees.
Then they swarm around me,
A hive of honey bees.

Who wouldn't like to have this effect on people? Can you learn how to get in the charisma zone? Absolutely, through changing how you move. Through improved movement your body feels spacious, fluid, loose and languid from the inside, while on the outside your movements are smooth, sensual, and visually compelling. As your mind and body become more congruent and connected you naturally are more charismatic and compelling to others. Cary Grant *learned* to become charismatic and charming. Much of his attraction he attributed to his early years working as an acrobat where he learned how to move with grace and balance. As he honed his acting skills it all came together to become Cary Grant, the persona men want to emulate and women wish more men were like.

Listening with compassion, empathy and attentiveness is a trademark of charismatic people. Many books on relationships discuss the importance of listening, but few offer concrete ways to cultivate it. It's all very well to have a desire to listen, but the practice of it is quite difficult when your mind is busy thinking about something else. It has been said that to listen is the ability to hear what is being said from the point of view of whoever is saying it. When I am not present my mind is busy making up its own story such that I miss relevant parts of conversations, which can quickly lead to misunderstandings. Too often we miss the context of what someone says because we are not present and aware in that moment. The next time you are in conversation try to breathe from your diaphragm as you interact. Make it a normal breath, not a big breath like a freight train, just feel how to reduce the effort of breathing. You may struggle to listen and focus on how you breathe at the same time, but with practice it will become easier. When we speak with one another the words we say can be directly correlated with what we feel as we say them. Just as we feel through our other senses, spoken language converges with body language in a timeless dance that truly reflects the "art" of conversation. In awareness listening and learning from one another is effortless, instantaneous, and greatly improves our ability to be flexible, inclusive, and expansive in our thoughts

Moving from your center, breathing from your diaphragm, yielding from within and without, with rotation and counter-rotation, fully differentiated, and propelling your body forward with gravity all lead to optimal function. You can look forward to excelling in any physical activity, feeling and looking younger, and improving stamina, power, strength, agility, flexibility and more. As your movements improve you think more clearly and authentically such that it becomes easier to distinguish fact and reality from fiction and belief. The path to congruence between what you feel and what you say and think is achieved through moving as your structure is organized. In discussing his art, the actor Kevin Spacey said, "Body language, what you feel and sense is always true, but what you say can be false." Becoming congruent with what you say and think and your body language means you can feel the consequences of your actions or words. If it feels bad, then you are less likely to say or do it.

The creative zone can only be reached if your mind is fully interconnected with your body through movement. In the creative zone you are optimally primed to allow the next big idea to enter into your consciousness. The creative zone is exemplified by James Clerk Maxwell--who with one brilliant theory unified electricity and magnetism, opening the path to the development of radio, television and microwaves--said as he lay dying, "What is done by what is called myself is, I feel, done by something greater than myself in me."[31] The powerful non-conscious processes within, and not some "god given talent," make it possible to function in the creative zone.

One constant in life is change. It is never too late to change, even though it may feel harder as you get older, but what is the alternative? There is no better moment than now to start your own transformation. As Darwin states so eloquently on evolution, "It is not the strongest who survive or the most intelligent, it is those who are most adaptable.", and *how we move* is the source of our adaptability.

10 - Eliminate Pain

Pain in movement is your brain telling you to
change how you move.

Globally over 98% of adult's experience pain from neuromuscular causes. Treating chronic pain is fast becoming one of the most expensive areas of medicine as costs have skyrocketed between $560 to $635 billion dollars per year.[32] The cost to business is enormous in terms of lost work and medical treatments for symptoms that rarely resolve the problem. Even though studies have shown the effectiveness of changing how we move to address neuromuscular pain the National Institute of Health (NIH) does not even list it as an option.[33] We are bombarded with a plethora of approaches that claim to resolve or relieve pain: drugs, surgeries, alternative therapies and supplements, physical therapy and exercise, pain management approaches, and more. It has become a multi-billion-dollar business across the spectrum making it very hard to determine what is truly effective.

Pain medications and muscle relaxers are designed to mask pain for a short time. Masking pain can be a welcome relief but it is also dangerous because you can do more damage as you unwittingly move in ways that are the cause of your pain. In addition, many drugs such as opioids are highly addictive and lead to further disconnect from what you feel. Some pain management approaches advocate learning to live with pain. In practice this often means you learn to ignore pain. By paying less attention to what you feel a cycle of poor movement layered on poor movement is perpetuated, which often leads to more chronic pain. Pain can overwhelm your consciousness so that you are unable to be aware of or have consideration for those around you. As many clients have said after relief from years of pain, "My life no longer revolves around my pain, I can take an interest in and be concerned about what is happening around me."

Another consequence of masking or ignoring pain is that the brain becomes so sensitized that any stimulation can trigger pain receptors called nociceptors. Once a nociceptor response has been activated, the neurons responsible become inflamed, leading to hypersensitivity. This means that a lesser stimulus in the same spot will activate a more severe pain response. Once a pain pathway is stimulated, even the smallest movement or touch can trigger it. I once saw a woman who screamed in pain when anyone touched her skin. Between the multiple drugs cocktail she was taking and her overstimulated pain response nociceptors it was impossible for me to help her. Thankfully, research is being done to develop therapies that disrupt the pain nociceptors.

Low back pain affects over 70% of the world's population making it the number one type of chronic pain. The usual medical approaches consist of medications, physical therapy, epidural injections, or back surgery. Some people do recover using the standard treatments, however evidence shows more people than ever do not. I could have become one of the victims of failed treatments but fortunately I got lucky and found an approach that worked. It worked because, unlike any other modality, it addressed the cause of my shoulder pain. Experience and observation of how most people move shows that *almost all neuromuscular pain is due to poorly organized movement.*

I saw a woman in her late eighties that had suffered from back pain for over fifty years. She was never a candidate for surgery because there was no medical injury to fix. Her condition was so debilitating that she could not stand for more than a few minutes before having to lie down. She had tried many different alternative modalities over the years but none offered lasting relief. After one session she informed me that she could feel it was helping. In just a few weeks she was consistently pain free and off her pain medications. She finally could spend time in her beloved garden, travel with her family, attend club meetings, cook, clean her house, and most importantly, be independent of daily caregivers. She gleefully told me about a trip with her family, walking around San Francisco all day pain free! All I did was show her how to move in ways that did not cause pain. No one should suffer needlessly when there is a scientifically sound, practical, intelligent approach to overcome neuromuscular pain.

The distraction of pain can contribute to impatience with others. During sessions to relieve her back pain, a client bitterly complained that she wanted to strangle her husband because he endlessly repeated the same questions all day long. As her pain lessened, she became bemused rather than openly annoyed by her husband's incessant questions. Finally out of pain she no longer reacted with anger and instead patiently answered his questions. Eventually he stopped the behavior. Getting out of pain can improve relationships as how we interact transitions to be more compassionate and empathetic.

The way government handles disability creates unintended dependencies in chronic pain sufferers. In the US, with a weak social safety net, people who cannot work, or who work less due to pain, suffer economic hardship which exacerbates and compounds their pain as it becomes inextricably linked with emotional and financial stress. In one instance, I saw a woman with severe carpal tunnel syndrome who had had two surgeries that did not resolve or reduce her pain, and she was facing yet another surgery. After just two sessions she was astounded by how much her pain had lessened, yet she felt she could not see me anymore because she was afraid. Since she had not worked for two years she was not sure she could handle it, but if she tried she would automatically lose her disability check. She chose to get the third surgery and continue receiving her disability check. Had she been allowed to continue her disability claim for some limited time while simultaneously working, taxpayers might have saved tens of thousands of dollars.

Pain leads to many other potentially costly health problems. One client, after sustaining injuries that eventually led to chronic pain, gained over one hundred pounds. He couldn't exercise or do any physical activities he had previously enjoyed. He became depressed and despondent. Eating offered a fleeting distraction and temporary comfort from the pain. He was beginning to suffer the consequences of obesity when he started seeing me. Over the course of the year I saw him he gradually began to lose weight. I asked if he was on a diet, but he said no, he was more active and food and eating had faded in importance. In discussing food he made an interesting observation: he can feel when he has eaten enough and he eats less junky food because his body feels bad when he does. He eventually lost over one hundred and fifty pounds[34] and the weight is still off many years later. Today he is physically fit and active, and is fully engaged in his work life making public presentations and more.

People in pain do not experience that rarefied space of effortless effort called the zone because getting in the zone *depends* on not being in pain. One client explained that when he meditated he "left" his body so that he could escape the discomforts of pain for a bit. There is a vast difference between not feeling your body because you learned to disconnect from it and being fully aware and present to how you move in any given moment. In the case of meditation, it is the difference between being meditative (or in the zone) throughout your day versus floating about in your mind for a short time. One allows you to be productive and creative in any given moment while the other offers temporary escape and relief from your body.

As we get older most of us feel it is inevitable that we become less agile and weaker, and we experience more pain. Stiffness can make you feel very old as any activity feels harder and you tire more easily. I play golf with older people who complain that their scores used to be in the low eighties, now they are lucky to be under one hundred. Is it inevitable that getting older renders you less skilled? Some aspects of aging are inevitable, but must technique be lost as well? Lost technical ability is almost always due to movement patterns that cause muscles to be in conflict with how your body is best organized to move. Getting older does not have to equate with losing function. As one client said, "When I used to wake up in the morning, it would take an hour or more before I felt like my body could move, and I was already tired! Now I don't feel stiff and sore, or ache when I get out of bed. My days are filled with activities that I never thought would be possible again. I feel so much younger and more vital." Focus on moving the way your body is best organized and you will maintain skill levels and potentially improve them.

We rely on many palliative remedies to temporarily relieve pain and discomfort, but the better solution is to address the underlying causes of pain and empower yourself to resolve it permanently. Getting out of pain restores a friendlier emotional connection to your body and interest in learning how to function better in every endeavor. As your relationship with your body improves you have the opportunity to become familiar with all the early warning signals so that you never should get to the point of pain before changing how you move. Your life will no longer revolve around fear of pain as you learn to move in better ways.

Pain resolution is not going to happen anytime soon through the medical system so it is up to you to take responsibility for addressing it on your own. Change to move according to the design of your skeleton, respond appropriately to the messages your brain is sending, and feel how you move in any given moment is the answer. Taking charge and committing to feel how to change your movements will resolve most pain conditions. Business can help by offering employees movement classes to keep their workforce physically healthy, functional, and productive. Many companies already provide services to employees such as massage and exercise classes, so adding movement classes and individual movement sessions to their offerings could be the most cost effective investment they can make to decrease lost work days, and as an added bonus, foster greater creativity and productivity.

11 - Brain Injury & Stroke Recovery

Recovery must occur from what you feel as
you move, not what you see.

Almost 800,000 people suffer from stroke every year in the US. Statistics show that just 10% of stroke victims ever recover full functionality. Health and caregiver costs associated with stroke run about $34 billion, including missed work days. Cerebral palsy affects 764,000 people in the US and 1 in 323 children are born with it worldwide. There are no statistics that track functional recovery of people diagnosed with cerebral palsy, but costs for caregiver support and medical care comes in at over $11 billion and counting. Traumatic Brain Injury (TBI) is estimated to be approximately 235,000 hospitalizations in the US per year, costing between $48 billion to $56 billion. Recovery from TBI is mixed depending on the severity of the injury. With recovery statistics as low as these it is in the best interest of everyone to find better ways to help individuals recover functionality and achieve greater independence.

Regaining arm functionality after a stroke or TBI can be a game changer. The best options available today are based on research by Edward Taub, who showed that after severing the sensory nerve in a monkey's arm, the monkey could still learn to use it if the "good" arm was constrained. Taub started a clinic using what he coined as "constraint induced movement therapy" (CIMT) to help people recover function of their affected side. The therapy works by restraining the "good" limb using a sling or mitt, thereby forcing the patient to use the affected limb. It has been quite successful in helping people regain some function; however, there are major recognized drawbacks:

- Even though people regain function they note that the affected limb does not "feel" as if it is part of them.

- The time involved and the cost of helping each patient is enormous, as they must be available for 6 to 9 hours per day, every day, for at least six weeks or more.

- The approach is not able to help patients learn refinements such as how to extend their fingers to open the hand.

Scientific research shows that in the first three months after a stroke the brain enters a growth phase of molecular, physiological, and structural change that in some ways resembles the brain environment of infancy and early childhood. It is a unique opportunity that can lead to great gains in functionality, but to take full advantage we must address how the brain evolved to learn. Relearning must take place through the senses, and as stated earlier, the senses are activated by movements. Regaining arm functionality starts with feeling the arm move. The best way to feel any movement is to close your eyes. Once eyes are closed a skilled practitioner can help an individual sense what it feels like to lift their lower arm, upper arm, hand, and learn to distinguish the difference between holding an arm up and releasing it down.

If we relearn a skill visually, without awareness of what it feels like, we create only a partial sensory map in the brain. For instance, a client with a severe brain injury from an auto accident recovered use of her right hand using constraint techniques, but it was not intrinsically part of her because if she lost her balance she did not automatically put the hand out to catch herself. One of the factors that make a brain injury so debilitating is that many people can see the arm or leg that is not working, but no matter how much they "will" it to move, it won't. Visual recognition of something does not equate to functional recovery. If I ask people with a brain injury to close their eyes and move a hand or finger, they find out they don't even feel where the hand or finger is, let alone *feel* how to move it! Visually identifying objects is a small part of functional recovery.

Converging and integrating all the senses to create a complex and dynamic internal sensory map is what leads to cognitive recovery. I saw a man for some time who had had a massive stroke, and one day after a session he got up, looked out the window and was suddenly aware of topography. Prior to this, he had great difficulty walking up and down hills because visually everything was flat. He did not see the contours of the landscape nor did he feel it in his body to adjust to different terrain. The same client had limited ability to read newspapers or books, and even have a casual conversation. His life was structured around simple tasks. He came in one day complaining that his mind was filled up with different thoughts. Some neurological connections were restored and his conscious mind was finally able to receive a plethora of information.

In recovering movement functionality the focus is often on range of motion and strength. However, strength and range of motion without the context of a task is quite irrelevant to the brain. One woman, after seven years of physical therapy for a stroke, complained to me that she learned to lift the arm above her head but she didn't know how to keep it there. She could grab an item with her hand but she didn't know how to let go of it. She had been a great knitter, but she couldn't get her fingers or hands to move with the control required to knit. When I asked her to tell me what it felt like to lift her arm or release it she could not tell the difference. She had no possibility of controlling her arm for a task when she didn't know what it felt like to lift or release the arm. Muscles tend to be flaccid or chronically contracted after a brain injury, so distinguishing between what it feels like to contract and release muscles is critical in recovery. To relearn a pathway of use muscles must be contracted until they register in the consciousness. Once felt it is possible to refine usage so that movements can become habitual and effortless. As important as it is to feel when muscles are contracting, one must also feel how to release muscles.

Neurological pathways are much like a highway system where we ideally navigate the most direct route to our destination. When we pile on patterns of movement that create circuitous routes to the same destination, we create a potential logjam. A stroke or brain injury literally causes us to lose some of the circuitry making it impossible to take the same meandering route. After his stroke, a client was having problems trying to relearn how to re-engage his flaccid left arm. Before his stroke the range of motion of his arm was quite limited. If the joints are restricted and limited it is difficult to feel how the arm should work. For example, the elbow moved in one hinge-like dimension. We had to re-establish full range of motion through the elbow joint *before* he could feel the upper arm separately from the lower arm. As we released the wrist joint he started to feel each finger separately from one another. The neurological pathways must be cleaned up and streamlined so the brain can re-establish appropriate neural connections. By cleaning up the sensory map, the more direct neurological routes are restored which leads to a better functional recovery in mind and body.

Multiple sclerosis, motor neurone disease (ALS) and post-polio syndrome are progressive conditions that can be slowed down by streamlining movements to become more functionally efficient. As connections die off, the brain can more easily make new connections to remain functional much longer. Other diseases, such as Parkinson's, which affects the nervous system by causing involuntary movements and tremors, can be reduced by learning to use just the muscles needed in an action.

By the time many clients see me they are usually in a deep depression at their lack of recovery. Helping them have success quickly is key to turning their mind around to one of hope and engagement. The ideal time to see people is right after a brain injury, before they habituate movements resulting from the event or disease, making it still more challenging to recover. Sometimes it is just too late, as it was with one woman. She could do an amazing number of movements, but she could not control any of them enough to do a task, such as washing her hair. Once her affected arm was on her head she couldn't feel how to move it, and because she had no idea when she was contracting or releasing muscles the arm would either fall down or contract, putting pressure on her head. She would become enraged and lash out at whoever was near. She had become so frustrated and depressed that she refused to do anything at home, such as taking her plate to the kitchen, even though she could. The emotional frustration of not knowing how to perform a simple task had led to anger and hopelessness.

To appear to other people that you have recovered functionality yet to know inside that you are far from it must be incredibly frustrating. Most people give up and become totally dependent on their caregivers, but as brain injury and other neurological diseases are happening at a younger age, it is even more imperative to help people recover functionality in a better way. Already under stress from lifestyle choices that have led to diabetes, heart disease, stroke and more, we will not always accept a medical system with limited treatments that often lead to economic hardship.

12 - Optimize Childhood Learning

How we move is the source of our intelligence.

Every healthy baby is born with equal possibility to be extraordinary. For a child to realize their full potential they must be allowed to learn as their brain evolved. Our system of education is based on the factory-model classroom which, at the turn of the nineteenth century, was deemed the most efficient way to rapidly scale a system to educate the masses. With limited classroom space and a single teacher, it was expedient to separate physical activities from academic studies. The problem with the separation is that science now knows the source of all information that enters the brain is *through movement*.[35] The classroom setting where children are expected to sit quietly in chairs for many hours defies our evolution. In addition, we did not evolve with the innate knowledge of how to sit in chairs. We evolved to squat such that weight bearing is through the feet. When sitting, weight is transferred to the sit bones, which effectively become our "feet".

Studies have shown that even a little physical activity before and during class increases students' ability to process and retain new material.[36] Studies conducted by Alejandro Lleras and Laura E. Thomas[37] show that movements related to a specific problem enhances the ability to solve it. Students who learn math using a tool such as an abacus[38] do better because, rather than solve problems "in their head" which requires conscious memorization, they use vision, touch, and more to expand their sensory map and reinforce memory. Some teaching methods, such as the Montessori system, teach reading and math using wooden blocks with letters and numbers on them. Students move spatially and visually to physically place the blocks in sequences and configurations to solve problems, thus improving their cognitive learning potential. The importance of movement is starting to be advocated in our educational systems but what is still not recognized is that *how we move is the source of optimal learning and cognition.*

The way we test and promote academic skills and sports is based on being competitive with one another which has led to isolation and adversity. In primatologist Frans de Waal's book "The bonobo and the atheist," he believes, based on studies with primates, that humans evolved to be social and highly empathic animals, where cooperation and support of each other is paramount to the survival of the group. Creating collaborative classrooms that encourage kids to cooperate and work with one another to successfully solve problems is far healthier as they learn social skills that will serve them well as adults.

Every activity a child does contains elements of every subject being taught. Some educators are no longer teaching by subject; they teach subjects within activities. Quest University in British Columbia Canada teaches using block plans that immerse students into an area of study such as biology. All realms of biology are discussed and students get further exploration through field trips.[39] Other approaches include concept based learning.[40] Students are encouraged to bring their experience and skills to real world concepts. Learning that is driven by overarching concepts necessitates that students transfer their knowledge between personal experiences, learning from other disciplines, and the broader global community. Cross-curricular[41] teaching approaches are becoming more popular to help teachers pool their talents to maximize coherence, relevance, and connections among content-areas being taught. The way we teach today isolates teachers amongst one another, and isolates students into test-taking memorizers, and it must stop.

Children with special needs

Children born with brain injury, neurological conditions, and loss of motor skills have the best chance of reaching their potential through movement learning. Evidence shows that even with minimal brain activity a child can learn an enormous amount of functionality--enough, at the very least, to achieve some degree of autonomy. Many children have the potential to achieve much greater independence and functionality despite medical prognosis. I saw a boy with mild cerebral palsy whose prognosis by his doctors was that he, due to his physical and cognitive limitations, would eventually need to be institutionalized. He was brought to see me at six years old to help him be more balanced in walking as the concern was that he was falling too often and would injure himself as he got older. I began by showing him how to better control each part of his skeleton for movement. At the end of the session I applied traction which uncompressed his spine and resulted in an instant growth spurt of five inches, as measured by his stunned doctors. In just a few sessions he was talking more, making eye contact, and playing with his older siblings and classmates. Within six months his teachers determined it was no longer necessary for him to be in special education and moved him over to regular, age appropriate classes. By the time he was seven he was excelling beyond his peers in all subjects. He learned to run, play soccer, dance, and do martial arts. His progress during the time I saw him was fully documented by staff at the medical institution, yet no one pursued or discussed whether the approach might work for other children or adults with the same condition or similar challenges. One little boy was very lucky. He is poised to live an independent and productive life, but what about the thousands of families that don't even know there are effective alternatives? Parents struggling with a limited system to help their special needs children gain functionality is costing society enormously. Many children, despite years of physical, speech, and other standard therapies, have limited movements and cognition such that they will eventually need to be in permanent care facilities. Many will be under institutional care until the end of their days, costing millions and tragically, their potential never realized.

Other evidence shows that babies born with severe brain damage can develop remarkable capabilities if purposeful movement is encouraged at birth. In a TV series called "The Brain" a child in japan who was born with only a quarter of his brain learned to see, talk, walk, and function independently despite dire predictions.[42] His mother spent a great deal of time touching and moving him in specific ways before sending him to a school that specialized in helping children develop cognition through movement learning.

Even if a child has damage to the part of the brain that reasons and thinks, perhaps the sensory part of the brain can take over some of that functionality since the thinking and sensory centers of the brain have their own separate memory stores. This is an intriguing area of study to determine what relationship, if any, sensory memory has to other types of memory. Is there a way to capitalize on it when other types of memory are damaged?

What if a child does not learn certain movements by a certain age? Is there such a thing as a developmental window? Neuroscientist Michael Merzenich conducted studies with cats which showed that, if one eye was covered during the first thirty days of its life, that eye never learned to see.[43] He concluded from the experiment that there is a window of opportunity for certain functionality that, if ignored, is gone forever. The first three years of a child's life are the most critical developmentally[44] when the "use it or lose it" rule likely applies to a wide swath of functionality.[45] I saw a child who, due to birth complications, doctors warned his parents not to let him sleep on his belly because of the danger of SIDS (sudden infant death syndrome). His parents were so alarmed and worried that they *never* let him be on his belly, day or night. The consequence of their well-intentioned fear is that he did not learn how to bear weight through his limbs. Critical to walking is learning to bear weight and propel the body using the constraint of a surface, such as the ground. Learning to bear weight through the body by pushing against a surface to move another part of the body is one of the earliest sensations we must experience, because it is the precursor to lifting the head. It is used in almost every subsequent movement we learn. The little boy I saw could do none of these things. His sitting up technique was a sit-up, not using his arms at all. He had no idea how to lean forward onto his hands to support himself in sitting or how to roll onto his belly and lean on his elbows. As much as I tried to help him feel how to bear weight through his arms to support himself he could not get it. The "use it or lose it" window of opportunity may have already passed. A child that learns functionality out of sync with the sequence of natural learning may lead to faulty and incomplete brain mapping. If a child learns a function, and then loses it due to an injury and loss of motor skills, is more likely to successfully relearn lost functionality versus a child that never learned it in the first place. I saw two children, both older than three years, who did not know how to swallow, suckle, see, or move with intention.

Much as I tried to help them form a purposeful movement, their body was wholly unknown and unknowable to their brain. They could not form an intention like grasping or touching something or even bringing food to their mouth. One child developed steadier eye movements but there was no comprehension or recognition when he looked at people. It was too late for the children I saw. They no longer had the capacity to become aware of their body from which to establish an intention, a logical or rational idea, or any purpose. Their window of opportunity for *any* cognitive learning was past.

The single most important change we can make for the continuance of the human species is to reform how we educate and prepare our children to be citizens of the world. Base education and teaching on how the brain evolved to learn through movement, and integrate the quality of how we move into the education curriculum. Our society will be better served and healthier in many ways by transforming how we approach learning and cognition of all children. It is a huge undertaking but we already know the consequences of not doing it.

13 - Embodied Spirituality

Balance is the key, effortless effort is the Way
—Adyashanti

Around the world there is a growing interest in spirituality as people search for more meaning in their life. Spirituality in religion based approaches include contemplation (like meditation) among its practices. Eastern philosophies that include practices such as yoga and meditation are a popular basis of spiritual approaches. The practice of contemplation and meditation can lead to a state of being that is meditative. Being in a meditative state is much like being in the zone—where all you feel, do, and think is effortless and timeless. In the zone, there are thousands of tiny movements and signals you catch when interacting with others, which informs how you respond and react. The meditative and zone state is fluidity, quiet from within, balance in mind and body, emotional clarity, rational thinking, truthfulness, and a sense of community and connection with all.

Many years ago I tried meditation, but the pain and stiffness in my body was such an overwhelming distraction I never had a hope of becoming meditative. I know I am not alone because most clients come to see me with the same complaint: body discomforts are interfering with their spiritual practices. As I became immersed in learning to move better, by working with myself and clients, I noticed that I entered a zone state more often. It is quite a relief that I can be meditative without needing to meditate, because frankly in a busy day with only so much time I would have to choose between going outside and running around versus sitting and meditating. Being outside with my horses, skiing, playing tennis, golfing, hiking, and so on is way preferable for me personally because my mind and body finds stillness in movement whereas being still makes my mind and body restless.

The body and the senses exist in the here and now, and they inform the mind of how it feels in any given moment as we go about our day. Through better connection to what you feel is the potential for the mind to be more present and aware. If your mind is disconnected from what you feel through movements it exists in a vacuum of its own creation, unable to be present or aware of others, regardless of your spiritual practice. One client said that he had learned to meditate by "leaving" his body. He had devised yet another form of disconnect leaving his mind ungrounded from his body, isolated and alone. Other movement approaches help to achieve some level of reconnect, but if you desire to function fully connected between your mind and body, it is the *quality of how you move* in any activity that matters.

By using the practices in this book to focus on how to move better, pain and discomfort are eliminated. As the body becomes better balanced all movements become easier. No longer distracted by bodily discomforts the mind is freed up to allow other sensations, thoughts, and ideas to enter. More optimal movement is integrated into every activity—yoga, meditation, washing dishes, walking, running, cleaning, and so on. Try the following practice to see if focusing on your movements brings you into a more present state of being:

- Lie on your back on the floor. Notice what parts of your body touch the floor and what is not touching. Notice how heavy your body feels. Do you feel stiff? Stiffen your legs in a stretch by pointing the toes and locking the knees. Hold while you breathe in and out a few times. Release, and then do it again, only this time imagine someone is pulling your toes down as you unlock your knees and release and lengthen your calves, upper legs, buttocks and belly. Did you hold your breath? Try again, breathing and releasing every muscle you can feel so that your sense is of your toes being pulled down.

- This time stretch your legs by lifting the toes. Notice how tight your hamstrings, buttocks and back get. Did you hold your breath again? Try it again, but imagine someone is pulling your heels down as you release all your muscles, allowing your legs to lengthen. Breathe as you repeat the movement. Repeat a few times feeling what to release getting the sense of lengthening as the heels are "pulled" down.

- Put your hands somewhere above your head so that your elbows and lower arms rest on the floor. Now stretch just your arms up without stretching the legs. Notice if they came off the floor. Try to keep them on the floor as you stretch. This time, imagine someone is pulling your fingers up as you release your elbows, shoulders, ribs, neck and jaw. Repeat several times releasing muscles you feel as you breathe. Notice if you eventually feel as if your ribs are coming apart from one another. Sense how to release them even further as you sense the pull.

- This time, imagine someone is pulling from your wrists to lengthen your arm. Release any muscles you feel to allow your ribs, chest, arms and shoulders to lengthen. Notice if your breathing contracts the belly, putting it in conflict with the lengthening. Feel how to breathe from the diaphragm to enhance the sense of lengthening. Keep allowing your neck and jaw to release as you feel the pull. Repeat with the opposite arm.

- Finally, lengthen the arms and legs letting them feel as if you are being pulled down by the toes and up by the wrists. Allow your entire body to lengthen as if you are growing an inch or two. Release every muscle you feel as you lengthen and breathe. Pause, and let your arms come down by your sides. Feel the back of your body and how it lies on the floor now. Notice whether more of it is touching the floor and whether it feels softer and lighter on the floor. Notice the quality of your flesh—is it hard or soft?

- Lie still for a couple of moments, noticing your neck, jaw, chest, back, legs, pelvis and breathing. Come to stand trying to find a way to do so without contracting your body. Make sure your head hangs forward and feel how you can move your body in ways that do not resist gravity, but instead flow with it. Once you are up, standing or sitting, allow your eyes to softly find your place in the book, pause and then continue reading for a couple of paragraphs. How aware of your body are you now? Is it easier to comprehend what you are reading? Are you more focused?

George Harrison of the Beatles wrote a song called <u>Within You Without You</u> that says the love you feel is what matters, because whatever your material success may be, without love within, your life is soulless. You cannot teach yourself to embody these qualities from your consciousness, they can only bubble through from what you feel as you move, non-consciously. If you cannot feel love within then how can you know what it feels like to love another? I replaced *seen* and *see* with *feel* in this lyric from the song to capture the essence of embodying the soul.

When you feel beyond yourself then you may find

Peace of mind is waiting there

And the time will come when you feel we're all one

And life flows on within you and without you

How we move is the path from within that helps us become aware of how our actions can lead us, one by one, to feel how to live again as we evolved; with compassion, cooperation, empathy, and love. When we embody our spirituality from within we thrive in harmony with all life.

About the Author

J. B. Mason lives in Northern California where there is ample opportunity to practice what she teaches. Living near Lake Tahoe she pursues many different outdoor activities such as skiing, hiking, running, mountain biking, horse riding, golf, tennis, and more.

Her most fervent desire is to help all society transform to collectively function according to the way the brain evolved. It is particularly important to transform medicine and health to create better outcomes for millions.

End Notes and Links

[1] https://en.wikipedia.org/wiki/Moshé_Feldenkrais

[2] https://www.youtube.com/watch?v=Phl82D57P58

[3] Learning to roll over http://www.pinterest.com/pin/182958803582521517/

[4] https://en.wikipedia.org/wiki/The_Description_of_the_Human_Body

[5] https://www.princeton.edu/main/news/archive/S35/82/65G58/

[6] www.animalspirit.org

[7] https://en.wikipedia.org/wiki/Sensory_system

[8] https://www.ted.com/talks/daniel_wolpert_the_real_reason_for_brains?language=en

[9] https://en.wikipedia.org/wiki/Evolution_of_the_brain

[10] "The User Illusion" by Tor Norretranders, page 142

[11] http://nautil.us/issue/10/mergers--acquisitions/trying-not-to-try

[12] http://movetoexcel.com/wp-content/uploads/2016/09/neuronal-plasticity.pdf

[13] http://www.ncbi.nlm.nih.gov/pubmed/19457048

[14] https://en.wikipedia.org/wiki/Artificial_neural_network

[15] http://www.scientificamerican.com/article/memories-may-not-live-in-neurons-synapses/

[16] Discovery Channel, "The Brain: Our Universe Within: Our Miraculous Mind" Kenji Kukue, Masakatsu Kakao

[17] http://nymag.com/scienceofus/2016/11/how-to-get-better-at-running.html

[18] http://well.blogs.nytimes.com/2013/04/03/reasons-not-to-stretch/?_r=0

[19] http://www.yourfamilyclinic.com/ND/si/si.html

[20] http://movetoexcel.com/wp-content/uploads/2016/11/guidelines.pdf

[21] Audio version: movetoexcel.com/public_html/wp-content/uploads/2016/10/howtofocus.ac3

[22] http://well.blogs.nytimes.com/2009/11/11/phys-ed-the-best-exercises-for-healthy-bones/?scp=1&sq=best%20exercises%20for%20healthy&st=cse&_r=0

[23] http://well.blogs.nytimes.com/2009/11/11/phys-ed-the-best-exercises-for-healthy-bones/?scp=1&sq=best%20exercises%20for%20healthy&st=cse&_r=0

[24] https://en.wikipedia.org/wiki/Johnny_Weissmuller

[25] http://www.cns.nyu.edu/home/ledoux/pdf/daed_LeDoux_2015.pdf

[26] https://www.youtube.com/edit?video_id=OdEHVLKRMV4

[27] http://www.physicsclassroom.com/class/newtlaws/Lesson-4/Newton-s-Third-Law

[28] Libet, Benjamin; Wright Jr., Elwood W.; Feinstein, Bertram; Pearl, Dennis K. (1979). "Subjective Referral of the Timing for a

Conscious Sensory Experience - A Functional Role for the Somatosensory Specific Projection System in Man". Brain **102**: 193–224.

[29] http://www.ux1.eiu.edu/~cfadd/1350/06CirMtn/NonuniformCM.html

[30] March of the cards: https://www.youtube.com/watch?v=aGqdhD0kMvw

[31] "The User Illusion" by Tor Norretranders, page 6

[32] https://www.sciencedaily.com/releases/2012/09/120911091100.htm

[33] http://www.ncbi.nlm.nih.gov/pubmed/1826550

[34] http://movetoexcel.com/testimonials/

[35] Sara Gable, Melissa Hunting (2001) "Nature, Nurture and Early Brain Development," University of Missouri Extension, http://extension.missouri.edu/p/GH6115

[36] http://www.dsr.wa.gov.au/brain-boost-sport-and-physical-activity-enhance-childrens-learning

[37] http://www.ndsu.edu/fileadmin/lthomas/Thomas_Lleras_2009a_.pdf

[38] https://www2.southeastern.edu/Academics/Faculty/dshwalb/files/Shwalb_et_alADPAbacusBookChapter.pdf

[39] https://questu.ca

[40] http://worldview.unc.edu/files/2013/07/Getting-the-Big-Idea-Handout.pdf

[41] https://www.edutopia.org/blog/cross-curricular-teaching-deeper-learning-ben-johnson

[42] Discovery Channel, "The Brain: Our Universe Within: Mind over Matter"

[43] http://merzenich.positscience.com/about-brain-plasticity/

[44] Kathy Sylva, *"Critical periods in childhood learning"* Department of Child Development and Primary Education, Institute of Education, London, UK, http://bmb.oxfordjournals.org/content/53/1/185.full.pdf

[45] http://bmb.oxfordjournals.org/content/53/1/185.long

Printed in Great Britain
by Amazon